MORE EVERYDAY SCIENCE MYSTERIES

STORIES FOR INQUIRY-BASED SCIENCE TEACHING

MORE EVERYDAY SCIENCE MYSTERIES

STORIES FOR INQUIRY-BASED SCIENCE TEACHING

Richard Konicek-Moran, Ed.D.
Professor Emeritus
University of Massachusetts
Amherst

Botanical illustrations by
Kathleen Konicek-Moran

National Science Teachers Association

National Science Teachers Association

Claire Reinburg, Director
Jennifer Horak, Managing Editor
Judy Cusick, Senior Editor
Andrew Cocke, Associate Editor
Betty Smith, Associate Editor

ART AND DESIGN
Will Thomas Jr., Director
Tim French, Cover and Interior Design

PRINTING AND PRODUCTION
Catherine Lorrain, Director
Jack Parker, Electronic Prepress Technician

NATIONAL SCIENCE TEACHERS ASSOCIATION
Francis Q. Eberle, PhD, Executive Director
David Beacom, Publisher

Library of Congress Cataloging-in-Publication Data

Konicek-Moran, Richard.
  More everyday science mysteries: stories for inquiry-based science teaching / by Richard Konicek-Moran.
      p. cm.
  Includes index.
  ISBN 978-1-933531-44-1
  1. Science—Methodology. 2. Problem solving. 3. Science—Study and teaching. 4. Science—Miscellanea. 5.
Detective and mystery stories. 6. Inquiry-based learning. I. Title.
  Q175.K6634 2009
  507.1—dc22
                        2009000956

MAY 06 2010

CONTENTS

The Stories and Background Material for Teachers

Matrix for Earth Systems Science and Technology

Matrix for Biological Sciences

Matrix for Physical Sciences

acknowledgments

This book is dedicated to my late parents, Pearl and Ervin Konicek, who helped and encouraged me all the way to be the first member of my family to earn college degrees. Without their support, I doubt if I would be writing this book right now.

I would like to thank the following teachers and administrators, who have helped me by field-testing the stories and ideas contained in this book over many years. These dedicated educators have helped me with their encouragement and constructive criticism:

Richard Haller
Jo Ann Hurley
Lore Knaus
Susan Johnson
Sharon Minor
Theresa Williamson
Third-grade team at Burgess Elementary, Sturbridge, MA
Second-grade team at Burgess Elementary, Sturbridge, MA
Alesia Peck
Teachers at Millbury Elementary Schools

I cannot express enough my thanks to the following educators, who have been my support over the plast 50 years in one way or another and helped me, through trials, to believe that these stories and inquiry really work in schools with real children:

Dr. Linda Denault
Dr. Klaus Schultz
Prof. Kathy Davis of the University of Massachusetts, Amherst

Former doctoral students and present colleagues:

Dr. Terez Waldoch
Dr. Diana Campbell
Helen Gibson, Science Coordinator for Holyoke, MA public schools
Dr. Betsy Koscher
Wanita Lafond
Barbara La Corte

My thanks also go out to all of the teachers and students in my graduate and undergraduate classes who wrote stories and tried them in their classes, as well as used my stories in their classes, and to the children at the Pottinger Elementary School in Springfield, Massachusetts, for their help in formulating many ideas that are found in this book.

To Professor Robert Barkman of Springfield College, who supported me and used the stories and techniques in workshops with Springfield elementary and middle school teachers, and to Marisa Creel, who was willing to be videotaped using her inquiry skills while teaching her class at the Pottinger Elementary School in Springfield, Massachusetts.

To my advisor at Columbia University, the late Professor Willard Jacobson, who made it possible for me to find my place in teacher education at the university level.

I also wish to thank Skip Snow, Lori Oberhofer, Jeff Kline and all of the biologists in the Everglades National Park, with whom I have had the pleasure of working for the past seven years, for helping me to remember how to be a scientist again. And to the interpretation group in the Everglades National Park, Katie Bliss, Bob DeGross, Maria Thompson, Laurie Humphry, and all of the other interpreters who helped me realize again that it's possible to help someone to look without telling them what to see and to help people form emotional connections with our resources.

My thanks to Bob Samples and Cheryl Charles for showing me another part of the creativity in education that I had forgotten was part of me.

My sincere thanks go to Claire Reinburg of NSTA, who had the faith in my work to publish the original book and this second volume; to Andrew Cocke, my editor, who helped me through the final steps; and to Tim French for his wonderful illustrations for the stories. In addition, I thank my talented wife Kathleen for her criticisms and for her beautiful botanical illustrations included in the background content sections of these stories. I also wish to thank Page Keeley and Joyce Tugel for their continued support of my work.

preface

"everyday miracles"

Questions often arise as to the origin of these everyday science mysteries. The answer is that they are most often derived from my everyday experiences. Science is all around us, and as we go through our daily routines it often eludes us, because as the old saying goes, "The hidden we seek, the obvious we ignore."

I am fortunate to be surrounded by a rural natural environment. My daily routine is predictable. I arise, eat breakfast, and then walk with my wife through the woods for a mile or so in order to exercise ourselves and our Australian shepherd and dachshund. They act as wonderful models as they exhibit their awareness of every scent and sight that might have changed over the previous 24 hours. Their noses are constantly sniffing the ground and the air in search of the variety of clues well beyond our limited senses. But as we walk, we look each day for our "miracle of the day." It may be a murder of crows harassing a barred owl or a red-tailed hawk flying over our heads with a squirrel in its talons. It might be a pair of wood ducks looking for a tree with a hole big enough for a nest, or a patch of spring trillium, or trout lilies. In the late summer, it could be a clump of ghostly Indian pipe and a rattlesnake plantain orchid in bloom or a hummingbird hovering near a flower. Sounds from the road bring questions about how sound travels and as we arrive home, we see crab apples, the worms in the compost pile, or the new greenhouse whose temperature fluctuations have plagued us all summer.

Textbooks are full of interesting information about the planets, space travel, plant reproduction, and animal behavior but offer very little about how this information was developed. Our world is full of questions, many of which are investigable by children and adults. Whenever possible, my senses and mind are drawn to these questions and stimulate the "I wonder…" section of my brain. I am intrigued by shadows, by the motion of the Sun and Moon during the daytime and the stars and planets at night. There are mysteries at every turn, if we keep our minds and eyes open to them. I am even more amazed that so many years have passed without my noticing so many of the questions that surround me. Writing these books has had a stimulating effect upon the way I look at the world. I thank my wife, a botanist, artist, and gardener, for spiking my awareness of things that I glossed over for so many years. We can get so caught up in the glitz of newsworthy science that we are blind to the little things that crawl at our feet, or sway in the branches over our heads, or move through the sky in predictable and fascinating ways each and every day. One can wonder where the wonder went in our lives as we get caught up in the search for better and better test scores. The stories spring forth by themselves when I can remember to see the world through child-like eyes. Perhaps, therein lies the secret to seeing those everyday science mysteries.

origins

In the mid-1990s, at the second Misconceptions Conference at Cornell University in Ithaca, New York, I remember having a conversation with Dr. James Shymansky, then at the University of Iowa, now at the University of Missouri in St. Louis, There, far above Cayuga's waters, we talked about an idea he espoused. The idea was for a new type of literature. He complained that current literature, for the most part, merely told children about what scientists had found out about certain phenomena and left out the drama of discovery and trial and failure. For this conversation and ideas gleaned from it, I offer heartfelt thanks.

Shymansky suggested that the current children's literature served a valuable purpose, but that it could also be written so that it offered a challenge to students, and that a skilled teacher could use such literature to parlay this challenge into classroom inquiry. He even offered examples of such literature, in the form of stories that would capture children's interests and leave the solution in their hands, rather than solve the challenge for them.

This encounter had a lasting effect upon me and I immediately went back to the University of Massachusetts and began to explore these possibilities with my students. At first I tried writing my own stories, and I tried out the idea with a seminar of my graduate students. We selected science topics and wrote stories about phenomena and added challenges by leaving the endings open, requiring the readers to engage in what we hoped would be actual inquiry in order to finish the story. We also added distracters of misconceptions, which were intended to double as formative assessments for the teachers.

Over the course of the semester we wrote many stories and the graduate students tried them out with the students in their classrooms. The children enjoyed the stories and we learned some important lessons on how to formulate the stories so that they provided the proper challenge.

For years after this initial experiment, I used the story concept with my graduate and undergraduate students in the elementary science methods classes. In lieu of the usual lesson plan, my class requirements included a final assignment that asked the students to write a story about a science phenomenon and include a follow-up paper that described how they would use the story to encourage inquiry learning in their classrooms. As I learned more about the concept, I was able to add techniques to my repertoire, which enhanced the quality of the stories and follow-up papers.

I learned that student teachers benefit from talking about their stories with other classmates and their instructor. They can gain valuable feedback before they launch into the final story. We organized small group meetings of no more than five students to preview and discuss ideas. We also designed a checklist document, which helped clarify the basic ideas behind the concept of the "challenge story." See sidebar.

As always, practice makes for a better product and eventually my students were producing stories that were useful for them and were acceptable to me as a form of assessment of their learning about teaching science.

As the years went by, teachers began to ask me if the stories I used as examples in class were available for them to use. They encouraged me to publish them in a book. So here is the second volume. I hope that it will provide you with ideas and inspiration to develop

Things to think about as you write your story

Does your story…

(1) address a single concept or conceptual scheme?

(2) address a topic of interest to your target age group?

(3) try to provide your audience with a problem they can solve through direct activity?

(4) require the students to become actively involved—hands-on, minds-on?

(5) have a really open-ended format?

(6) provide enough information for the students to identify and attack the problem?

(7) consider whether materials you intend for the students to use are readily available to them?

(8) provide opportunities for students to discuss the story and come up with a plan for finding some answers?

(9) make data collection and analysis of those data a necessity?

(10) provide some way for you to assess what their current preconceptions are about the topic? (This can be implicit or explicit.)

more inquiry-oriented lessons in your classrooms. And perhaps you may be motivated to try writing your own stories for teaching those concepts you find most difficult to teach.

I would like now to take you, in the introduction, to two classrooms where one story from volume one of *Everyday Science Mysteries* was used, and in the form of case studies, let you see how two teachers worked this story into their curriculum. You will note that both teachers continued data gathering for the full school year while working other units into their schedule and returned to the featured unit when they felt it was appropriate.

INTRODUCTION

CASE STUDIES ON HOW TO USE THE STORIES IN THE CLASSROOM

To open the book, I would like to introduce you to one of the stories from the first volume of *Everyday Science Mysteries* and then show how the story was used by two teachers, Teresa, a second-grade teacher, and Lore, a fifth-grade teacher. Then I will explain the philosophy and organization of the book before going to the stories and background material. Here is the story, *Where Are the Acorns?*

WHERE ARE THE ACORNS?

Cheeks looked out from her nest of leaves, high in the oak tree above the Anderson family's backyard. It was early morning and the fog lay like a cotton quilt on the valley. Cheeks stretched her beautiful grey, furry body and looked about the nest. She felt the warm August morning air, fluffed up her big grey bushy tail and shook it. Cheeks was named by the Andersons since she always seemed to have her cheeks full of acorns as she wandered and scurried about the yard.

"I have work to do today!" she thought and imagined the fat acorns to be gathered and stored for the coming of the cold times.

Now the tough part for Cheeks was not gathering the fruits of the Oak trees. There were plenty of trees and more than enough acorns for all of the grey squirrels who lived about the yard. No, the problem was finding them later on when the air was cold and the white stuff might be covering the lawn. Cheeks had a very good smeller and could sometimes smell the acorns she had buried earlier. But not always. She needed a way to remember where she had dug the holes and buried the acorns. Cheeks also had a very small memory and the yard was very big. Remembering all of these holes she had dug was too much for her little brain.

The Sun had by now risen in the east and Cheeks scurried down the tree to begin gathering and eating. She also had to make herself fat so that she would be warm and not hungry on long cold days and nights when there might be little to eat.

"What to do... what to do?" she thought as she wiggled and waved her tail. Then she saw it! A dark patch on the lawn. It was where the Sun did not shine. It had a shape and two ends. One end started where the tree trunk met the ground. The other end was lying on the ground a little ways from the trunk. "I know," she thought. "I'll bury my acorn out here in the yard, at the end of the dark shape, and in the cold times, I'll just come back here and dig it up!!! Brilliant, Cheeks," she thought to herself and began to gather and dig.

On the next day she tried another dark shape and did the same thing. Then she ran about for weeks and gathered acorns to put in the ground. She was set for the cold times for sure!!

Months passed and the white stuff covered the ground and trees. Cheeks spent more time curled up in her home in the tree. Then one bright crisp morning, just as the Sun was lighting the sky, she looked down and saw the dark spots, brightly dark against the white ground. Suddenly she had a great appetite for a nice juicy acorn. "Oh yes," she thought. "It is time to get some of the those acorns I buried at the tip of the dark shapes."

She scampered down the tree and raced across the yard to the tip of the dark shape. As she ran, she tossed little clumps of white stuff into the air and they floated back onto the ground. "I'm so smart," she thought to herself. "I know just where the acorns are." She did seem to feel that she was a bit closer to the edge of the woods than she remembered but her memory was small and she ignored the feelings. Then she reached the end of the dark shape and began to dig and dig and dig!

And she dug and she dug and she dug! Nothing!! "Maybe I buried them a bit deeper," she thought, a bit out of breath. So she dug deeper and deeper and still, nothing. She tried digging at the tip of another of the dark shapes and again found nothing. "But I know I put them here," she cried. "Where could they be?" She was angry and confused. Did other squirrels dig them up? That was not fair. Did they just disappear? What about the dark shapes?

How can she find the acorns? Where in the world are they? Can you help her find the place where she buried them? Please help, because she is getting very hungry!

How Two Teachers Used "Where are the acorns?"

Teresa, a veteran second-grade teacher

Teresa usually begins the school year with a unit on fall and change. This year she looked at the National Science Education Standards (NSES) and decided that a unit on the sky and cyclic changes would be in order. Since shadows were something that the children often noticed and included in playground games (shadow tag), Teresa thought using the story of "Cheeks" the squirrel would be appropriate.

To begin, she felt that it was extremely important to know what the children already knew about the Sun and the shadows cast from objects. She wanted to know what kind of knowledge they shared with Cheeks and what kind of knowledge they had that the story's hero did not have. She arranged the children in a circle so that they could see each other and hear each other's comments. Teresa read the story to them, stopping along the way to see that they knew that Cheeks had made the decision on where to bury the acorns during the late summer and that the squirrel was looking for her buried food during the winter. She asked them to tell her what they thought they knew about the shadows that Cheeks had seen. She labeled a piece of chart paper, "Our best ideas so far." As they told her what they "knew," she recorded their statements in their own words:

"Shadows change every day."
"Shadows are longer in winter"
"Shadows are shorter in winter"
"Shadows get longer every day"
"Shadows get shorter every day"
"Shadows don't change at all."
"Shadows aren't out every day."
"Shadows move when you move."

She asked the students if it was okay to add a word or two to each of their statements so they could test them out. She turned their statements into questions and the list then looked like this:

"Do shadows change every day?"
"Are shadows longer in winter?"
"Are shadows shorter in winter?"
"Do shadows get longer every day?"
"Do shadows get shorter every day?"
"Do shadows change at all?"
"Are shadows out every day?"
"Do shadows move when you move?

Teresa focused the class on the questions that could help solve Cheeks' dilemma. The children picked "Are shadows longer or shorter in the winter?" and "Do shadows change at all?" The children were asked to make predictions based on their experience. Some said that the shadows would get longer as we moved toward winter and some predicted the opposite. Even though there was a question as to whether they would change at all, they agree unanimously that there would probably be some change over time. If they could get data to support that there was change, that questions would be removed from the chart.

Now the class had to find a way to answer their questions and test predictions. Teresa helped them talk about fair tests and asked them how they might go about answering the questions. They agreed almost at once that they should measure the shadow of a tree each day and write it down and should use the same tree and measure the shadow every day at the same time. They weren't sure why time was important except that they said they wanted to make sure everything was fair. Even though data about all of the questions would be useful, Teresa thought that at this stage, looking for more than one type of data might be overwhelming for her children.

Teresa checked the terrain outside and realized that the shadows of most trees might get so long during the winter months that they would touch one of the buildings and become difficult to measure. That could be a learning experience but at the same time it would frustrate the children to have their investigation ruined after months of work. She decided to try to convince the children to use an artificial "tree" that was small enough to avoid our concern. To her surprise, there was no objection to substituting an artificial tree since, "If we measured that same tree every day, it would still be fair." She made a tree out of a dowel that was about 15 cm tall, and the children insisted that they glue a triangle on the top to make it look more like a tree.

The class went outside as a group and chose a spot where the sun shone without obstruction and took a measurement. Teresa was concerned that her students were not yet adept at using rulers and tape measures so she had the children measure the length of the shadow from the base of the tree to its tip with a piece of yarn and then glued that yarn onto a wall chart above the date when the measurement was taken. The children were delighted with this.

For the first week, teams of three went out and took daily measurements. By the end of the week, Teresa noted that the day-to-day differences were so small that perhaps they should consider taking a measurement once a week. This worked much better, as the chart was less "busy" but still showed any important changes that might happen.

As the weeks progressed, it became evident that the shadow was indeed getting longer each week. Teresa talked with the students about what would make a shadow get longer and armed with flashlights, the children were able to make longer shadows of pencils by lowering the flashlight. The Sun must be getting lower too if this was the case, and this observation was added to the chart of questions. Later, Teresa wished that she had asked the children to keep individual science notebooks so that she could have been more aware of how each individual child was viewing the experiment.

The yarn chart showed the data clearly and the only question seemed to be, "How long will the shadow get?" Teresa revisited the Cheeks story and the children were able to point out that Cheeks' acorns were probably much closer to the tree than the winter shadows indicated. Teresa went on with another unit on fall changes and each week added another piece of yarn to the chart. She was relieved that she could carry on two science units at once and still capture the children's interest about the investigation each week after the measurement. After winter break, there was great excitement when the shadow began getting shorter. The shortening actually began at winter solstice around December 21, but the children were on break until after New Years. Now, the questions became, "Will it keep getting shorter? For how long?" Winter passed and spring came and finally the end of the school year was approaching. Each week, the measurements were taken and each week a discussion was held on the meaning of the data. The chart was full of yarn strips and the pattern was obvious. The fall of last year had produced longer and longer shadow measurements until the New Year and then the shadows had begun to get shorter. "How short will they get?" and "Will they get down to nothing?" questions were added to the chart. During the last week of school, they talked about their conclusions and the children were convinced that the Sun was lower and cast longer shadows during the fall to winter time and that after the new year, the Sun got higher in the sky and made the shadows shorter. They were also aware that the seasons were changing and that the higher Sun seemed to mean warmer weather and trees producing leaves. The students were ready to think about seasonal changes in the sky and relate them to seasonal cycles. At least Teresa thought they were.

On the final meeting day in June, she asked her students what they thought the shadows would look like next September. After a great deal of thinking, they agreed that since the shadows were getting so short, that by next September, they would be gone or so short that they would be hard to measure. Oh my!! The idea of a cycle had escaped them, and no wonder, since it hadn't really been discussed. The obvious extrapolation of the chart would indicate that the trend of shorter shadows would continue. Teresa knew that she would not have a chance to continue the investigation next September but she might talk to the third-grade team and see if they would at least carry it on for a few weeks so that the children could see the repeat of the previous September, data. Then the students might be ready to think more about seasonal changes and certainly their experience would be useful in the upper grades, where seasons and the reasons for seasons would become a curricular issue. Despite these shortcomings, it was a marvelous experience, and the children were given a great opportunity to design an investigation and collect data to answer their questions about the squirrel story at a level appropriate to their development. Teresa felt that the children had had an opportunity to carry out a long-term investigation, gather data, and come up with conclusions along the way about Cheek's dilemma. She felt also that the standard had been partially met or at least was in progress. She would talk with the third-grade team about that.

Lore (pronounced Laurie), a veteran fifth-grade teacher
In September while working in the school, I had gone to Lore's fifth-grade class for advice. I read them the Cheeks story and asked them at which grade they thought it would be most appropriate. They agreed that it would most likely fly best at second grade. It seemed, with their advice, that Teresa's decision to use it there was a good one.

However, about a week after Teresa began to use the story, I received a note from Lore, telling me that her students were asking her all sorts of questions about shadows, the Sun, and the seasons, and could I help. Despite their insistence that the story belonged in the second grade, the fifth graders were intrigued enough by the story to begin asking questions about shadows. We now had two classes interested in Cheeks' dilemma but at two different developmental levels. The fifth graders were asking questions about daily shadows, direction of shadows, and seasonal shadows and they were asking, "Why is this happening?" Lore wanted to use an inquiry approach to help them find answers to their questions but needed help. Even though the Cheeks story had opened the door to their curiosity, we agreed that perhaps a story about a pirate burying treasure in the same way Cheeks had buried acorns might be better suited to the fifth-grade interests in the future.

Lore looked at the NSES for her grade level and saw that they called for observing and describing the Sun's location and movements and studying natural objects in the sky and their patterns of movement. But the students' questions, we felt, should lead the investigations. Lore was intrigued by the 5E approach to enquiry (*engage, elicit, explore, explain, and evaluate*) so since the students were already "engaged," she added the "elicit" phase to find out what her students already knew. (The five Es will be defined in context as this vignette evolves). So, Lore started her next class asking the class what they "knew" about the shadows that Cheeks used and what caused them. The students stated:
"Shadows are long in the morning, short at mid-day and longer again in the afternoon."
"There is no shadow at noon because the Sun is directly overhead."
"Shadows are in the same place every day so we can tell time by them."
"Shadows are shorter in the summer than in the winter."
"You can put a stick in the ground and tell time by its shadow."

Just as Teresa had done, Lore changed these statements to questions, and they entered the "exploration" phase of the 5E inquiry method.

Luckily, Lore's room opened out onto a grassy area that was always open to the Sun. The students made boards that were 30 cm square and drilled holes in the middle and put a toothpick in the hole. They attached paper to the boards and drew shadow lines every half hour on the paper. They brought them in each afternoon and discussed their results. There were many discussions about whether or not it made a difference where they placed their boards from day to day.

They were gathering so much data that it was becoming cumbersome. One student suggested that they use overhead transparencies to record shadow data and then overlay them to see what kind of changes occurred. Everyone agreed that it was a great idea.

Lore introduced the class to the *Old Farmer's Almanac* and the tables of sunsets, sunrises, and lengths of days. This led to an exciting activity one day that involved math. Lore asked them to look at the sunrise time and sunset time on one given day and to calculate the length of the daytime sun hours. Calculations went on for a good ten minutes and Lore asked each group to demonstrate how they had calculated the time to the class. There must have been at least six different methods used and most of them came up with a common answer. The students were amazed that so many different methods could produce the same answer. They also agreed that several of the methods were more efficient than others and finally agreed that using a 24-hour clock method was the easiest. Lore was ecstatic that they had created so many methods and was convinced that their understanding of time was enhanced by this revelation.

This also showed that children are capable of metacognition—thinking about their thinking. Research (Metz 1995) tells us that elementary students are not astute at thinking about the way they reason but that they can learn to do so through practice and encouragement. Metacognition is important if students are to engage in inquiry. They need to understand how they process information and how they learn. In this particular instance, Lore had the children explain how they came to their solution for the length of day problem so that they could be more aware of how they went about solving the challenge. Students can also learn about their thinking processes from peers who are more likely to be at the same developmental level. Discussions in small groups or as an entire class can provide opportunities for the teacher to probe for more depth in student explanations. The teacher can ask the students who explain their technique to be more specific about how they used their thought processes: dead ends as well as successes. Students can also learn more about their metacognitive processes by writing in their notebooks about how they thought through their problem and found a solution. Talking about their thinking or explaining their methods of problem solving in writing can lead to a better understanding of how they can use reasoning skills better in future situations.

I should mention here that Lore went on to teach other units in science while the students continued to gather their data. She would come back to the unit periodically for a day or two so the children could process their findings. After a few months, the students were ready to get some help in finding a model that explained their data. Lore gave them globes and clay so that they could place their observers at their latitude on the globe. They used flashlights to replicate their findings. Since all globes are automatically tilted at a 23.5-degree angle it raised the question as to why globes were made that way. It was time for the "explanation" part of the lesson and Lore helped them see how the tilt of the Earth could help them make sense of their experiences with the shadows and the Sun's apparent motion in the sky.

The students made posters explaining how the seasons could be explained by the tilt of the Earth and the Earth's revolution around the Sun each year. They had "evaluated" their understanding and "extended" it beyond their experience. It was, Lore agreed, a very successful "6E" experience. It had included the engage and elicit, explore, explain, evaluate, and the added extend phase.

references

Metz, K. E. 1995. Reassessment of developmental constraints on children's science instruction. *Review of Educational Research* 65 (2): 93–127.

Yankee Publishing. *The old farmer's almanac,* published yearly since 1792. Dublin, NH: Yankee Publishing.

CHAPTER 1

THEORY BEHIND THE BOOK

We have all heard people refer to any activity that takes place in a science lesson as an "experiment." Yet, as taught today, science is practically devoid of true experiments. Experiments by definition test hypotheses, which are also usually absent from school science. A hypothesis is a human creation developed by a person who has been immersed in a problem for a sufficient amount of time to feel the need to propose an explanation for the event or situation over which he or she has been puzzled.

However, it is quite common and proper for us to investigate our questions without proper hypotheses. Investigations can be carried out as "fair tests," which are possibly more appropriate for elementary classrooms, because children often have not had the experience of prior research to set up a hypothesis in the true scientific mode. I recently asked a fourth grade girl what a "fair experiment" was, and she replied that "it is an experiment where the answer is the one I expected." We cannot assume even at fourth grade that children know about controlling variables; it needs repeating.

A hypothesis is more than a guess. A hypothesis will most often have an "if…then…" statement in it. For example, "*If* I stand further away from a mirror, *then* I will see more of myself in the mirror." Predictions in school science should also be more than mere guesses or hunches, however. Predictions should be based upon experience and thoughtful consideration. Regularly asking children to give reasons for their predictions is

a good way to help them see the difference between guessing and predicting.

Two elements are often missing in most school science curricula: *sufficient time* to puzzle over problems and problems that have some *real-life application*. It is much more likely that students will "cover," in a prescribed time period, an area of study, say, pond life, with readings, demonstrations, and a field trip to a pond with an expert, topped off with individual or group reports on various pond animals and plants, complete with shoebox dioramas and giant posters. Or there may be a study of the solar system with reports on facts about the planets, complete with dioramas and culminating with a class model of the solar system hung from the ceiling.

These lessons are naturally fun to do, but the problem is that they seldom pose any real problems, nothing into which the students can sink their collective teeth, use their minds, ponder, puzzle, hypothesize, and then experiment.

You have certainly noticed that most science curricula have a series of "critical" activities in which students participate and which supposedly lead to an understanding of a particular concept. In most cases, there is an assumption that students share a common view or a common set of preconceptions about the concept so that the activities will move the students collectively from one point to another, hopefully closer to the accepted scientific view. This is a particularly dangerous assumption since research shows us that students enter into learning situations with a variety of preconceptions. These preconceptions are not only well ingrained in the students' minds but are exceptionally resistant to change. Going through the series of prescribed activities will have little meaning to students who have brought to the lessons conceptions that have little connection to the planned lessons.

Bonny Shapiro, in her book, *What Children Bring to Light* (1994), points out in indisputable detail how a well-meaning science teacher ran his students through a series of activities on the nature of light without knowing that the students in the class all shared the misconception that seeing any object originates in the eye of the viewer and not from the reflection of light from an object into the eye. The activities were, for all intents and purposes, wasted, although the students had "solved the teacher" (rather than the problems) to the extent that they were able to fill in the worksheets and pass the test at the end of the unit—all the while doubting the critical concept that light reflecting from object to eye was the paramount fact and meaning of the act of seeing. "Solving the teacher" means that the children have learned a teacher's mannerisms, techniques, speech patterns, and perhaps teaching methods to the point that they can predict exactly what the teacher wants, what pleases or annoys the teacher, and how they can perform so that the teacher believes they have learned and understood what she expected of them.

Eleanor Duckworth, in her monograph *Inventing Density* (1986), says, "The critical experiments themselves cannot impose their own meanings. One has to have done a major part of the work already, one has to have developed a network of ideas in which to imbed the experiments." This may be the most important quote in this book.

So, how does a teacher make sure that her students develop a network of ideas in which to embed such activities? How does the teacher uncover student preconceptions about the topic to be studied? I believe that this book can offer some answers to these questions and offer some suggestions for remedying the problems mentioned above.

WHaT IS INQUIrY aNYWaY?

There is probably no one definition of "teaching for inquiry," but at this time the acknowledged authorities on this topic are the National Research Council (NRC) and the American Association for the Advancement of Science (AAAS). After all, they are respectively the authors of the *National Science Education Standards* and the *Benchmarks for Science Literacy,* upon which most states have based their curriculum standards. For this reason, I will use their definition, which I will follow throughout the book. The NRC, in *Inquiry and the National Science Education Standards: A Guide for Teaching and Learning* (2000), says that in order for real inquiry to take place in the classroom, the following five essentials must occur. They are

(1) learner engages in scientifically oriented questions;
(2) learner gives priority to evidence in responding to questions;
(3) learner formulates explanations from evidence;
(4) learner connects explanations to scientific knowledge; and
(5) learner communicates and justifies explanations. (p. 29)

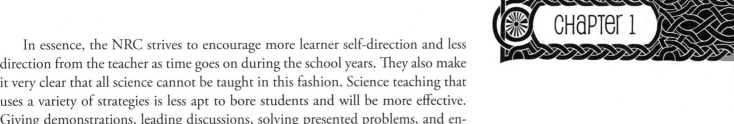

In essence, the NRC strives to encourage more learner self-direction and less direction from the teacher as time goes on during the school years. They also make it very clear that all science cannot be taught in this fashion. Science teaching that uses a variety of strategies is less apt to bore students and will be more effective. Giving demonstrations, leading discussions, solving presented problems, and entering into productive discourses about science are all viable alternatives. However, the NRC does suggest that certain common components should be shared by whichever instructional model is used:

(1) Students engage with a scientific question, event, or phenomenon. This connects with what they already know, creates dissonance with their own ideas, and/or motivates them to learn more.
(2) Students explore ideas though hands-on experiences, formulate and test hypotheses, solve problems, and create explanations for what they observe.
(3) Students analyze and interpret data, synthesize their ideas, build models, and clarify concepts and explanations with teachers and other sources of scientific knowledge.
(4) Students extend their new understanding and abilities and apply what they have learned to new situations.
(5) Students, with their teachers, review and assess what they have learned and how they have learned it. (p. 35)

THE REASONS FOR THIS BOOK

According to a summary of current thinking in science education in the journal *Science Education,* "one result seems to be consistently demonstrated: students leave science classes with more positive attitudes about science (and their concepts of themselves as science participants) when they learn science through inductive, hands-on techniques in classrooms where they're encouraged by a caring adult and allowed to process the information they have learned with their peers" (1993).

This book and particularly the stories within it provide an opportunity for students to take ownership of their learning and, as stated in the quotation above, learn science in a way that will give them a more positive attitude about science and to process their learning with their classmates and teachers. Used as intended, the stories will require group discussions, hands-on, minds-on techniques, and a caring adult.

THE STORIES

These stories are similar to mystery tales but purposely lack the final chapter where the clever sleuth finally solves the mystery and tells the readers not only "whodunit," but how he or she knew. Because of the design of the tales in this book, the students are challenged to become the sleuths and come up with likely "suspects" (the hypotheses or predictions) and carry out investigations (the experiments or

investigations) to find out "whodunit" (the results). In other words, they write the final ending or perhaps multiple possible endings. They are placed in a situation where they develop, from the beginning, "the network of ideas in which to imbed activities," as Duckworth suggests (1986, p. 39). The students are also the designers of the activities and therefore have invested themselves in finding the outcomes that make sense to them. I want them to have solved the problem rather than having solved the teacher. I do want to reemphasize, however, that we should all be aware that successful students do spend energy in solving their teachers.

In one story ("Color Thieves"), Jenny is excited about making an attractive and colorful cover for her science report. She is enticed by her mother's offer of using a colored transparent folder. She discovers that many of the colors on her cover look black and others disappear when viewed through the red filter of the folder. She is incensed that the red cover actually "stole" her colors and wants to know why. Students then explore the nature of light, color, and the spectrum, perhaps by writing secret messages, visible only through colored filters or by looking through filtered glasses. From these activities students learn about how we see, and correct any misconceptions about light and reflection. Truly this is science as process and product. It also means that the students "own" the problem. This is what we mean by "hands-on, minds-on" science instruction. The teachers' belief in the ability of their students to own the questions and to carry out the experiments to reach conclusions is paramount to the process. Each story has suggestions as to how the teachers can move from the story reading to the development of the problems, the development of the hypotheses, and eventually the experiments that will help their students come to conclusions.

Learning science through inquiry is a primary principle in education today. You might well ask, "Instead of what?" Well, instead of learning science as a static or unchanging set of facts, ideas, and principles without any attention being paid to how these ideas and principles were developed. Obviously, we cannot expect our students to discover all of the current scientific models and concepts. We do however, expect them to appreciate the processes through which the principles are attained and verified. We also want students to see that science includes more than just what happens in a classroom; that the everyday happenings of their lives are connected to science. Gathering flowers for observation, cooking pasta in a hurry, or sweetening iced tea are only some of the examples of everyday life connected to science as a way of thinking and as a way of constructing new understandings about our world.

There are 15 stories in this book, each focused on a particular conceptual area, such as thermodynamics, mirrors and reflection, light and color, temperature and energy, flowers, and decay and decomposition. Each story can either be photocopied and distributed to students to read and discuss in small groups or be read aloud to students and discussed by the entire class. During the discussion, it is the role of the teacher to help the students identify the problem or problems and then design ways to solve these problems.

Most stories also include a few "distracters," also known as common misconceptions or alternative conceptions. The distracters are usually placed in the stories as opinions voiced by the characters who discuss the problematic situation. For example, in "Cool it, Dude!" friends argue about the cooling times and displace-

ment when using crushed ice or ice cubes, voicing their opinions about the question at hand. These opinions include many of the misconceptions we know to be prevalent in our population, shared by both children and adults. Where do these common misconceptions come from and how do they arise?

Development of Mental Models

Until recently, educational practice has operated under the impression that children and adults come to any new learning situation without the benefit of prior ideas connected to the new situation. But research shows that in almost every circumstance, learners have developed models in their minds to explain many of the everyday experiences they have encountered (Bransford, Brown, and Cocking 1999; Watson and Konicek 1990; Osborne and Fryberg 1985). Everyone has had experience with differences in temperature as they place their hands on various objects; everyone has seen objects in motion and certainly has been in motion, either in a car, in a plane, or on a bicycle; everyone has experienced forces in action, upon objects or upon themselves. Finally, everyone has been seduced into developing a satisfactory way to explain these experiences, to have developed a mental model. It is also possible that individuals have read books or watched programs on TV and used these presented images and ideas to embellish their models. It is even more likely that they have been in classrooms where these ideas have been discussed by a teacher or by other students. In the film *A Private Universe* (Schneps 1986), it was shown that almost all of the interviewed graduates and faculty of Harvard University showed some misunderstanding for either the reasons for the seasons or the reasons for the phases of the moon. Many had taken high-level science courses either in high school or at the university.

According to the dominant and current learning theory called constructivism, all of life's experiences are integrated into person's mind; they are accepted or rejected or even modified to fit existing models residing in that person's mind. Then, these models are used and tested for their usefulness in predicting outcomes experienced in the environment. If a model works, it is accepted as a plausible explanation; if not, it is modified until it does fit the situations one experiences. Regardless, these models are present in everyone's mind and brought to consciousness when new ideas are encountered. Rarely, they may be in tune with current scientific thinking, but more often they are "common sense science" and not clearly consistent with current scientific beliefs.

One of the reasons for this is that scientific ideas are often counterintuitive to everyday thinking. For example, when you place your hand on a piece of metal in a room, it feels cool to your touch. When you place your hand on a piece of wood in the same room, it feels warmer to the touch. Many people will deduce that the temperature of the metal is cooler than that of the wood. Yet, if the objects have been in the same room for any length of time, their temperatures will be equal. It turns out that when you place your hand on metal, it conducts heat out of your hand quickly, thus giving the impression that it is cold. The wood does not conduct heat as rapidly as the metal and therefore "feels" warmer than the metal. In other words, our senses have fooled us into thinking that instead of everything in the room being at room temperature, the metal is cooler than anything else. Therefore our erroneous conclusion: Metal objects are always cooler than other objects in a room. Indeed, if you go from room to room and touch many objects, your idea is reinforced and becomes more and more resistant to change.

These ideas are called by many names: misconceptions, prior conceptions, children's thinking, or common sense ideas. They all have two things in common: They are usually firmly embedded in the mind, and they are highly resistant to change. Finally, if allowed to remain unchallenged, they will dominate a student's thinking, for example, about heat transfer, to the point that the scientific explanation will be rejected completely regardless of the method by which it is presented.

WHY STOrIES?

Primarily, stories are one of the most effective ways to get someone's attention. Stories have been used since the beginning of recorded history and probably long before that. Myths, epics, oral histories, ballads, and such have enabled humankind to pass on the culture of one generation to the next, and the next, *ad infinitum*. Anyone who has witnessed story time in classrooms, in libraries, or at bedtime knows the magic held in well-written, well-told tales. They have beginnings, middles, and ends.

These stories begin like many familiar tales do: in homes, in classrooms, with children interacting with each other or with arguing siblings, with classmates and friends talking, or with parents or other adults in family situations. Sometimes the characters are animals in the backyard or forest that are given personalities and engage in discoveries and problem situations of their own. But here the resemblance ends between our stories and traditional ones.

Science stories normally have a theme or a scientific topic that unfolds, giving a myriad of facts, principles, and perhaps a set of illustrations or photographs, which try to explain to a child the current understanding about the given topic. For years science books have been written as reviews of what science has constructed to the present. These books have their place in education, even though children often get the impression from them that the information they have just read appeared magically as scientists went about their work and "discovered" the truths and facts depicted in those pages. But as Martin and Miller (1990) put it: "The scientist seeks more than isolated facts from nature. The scientist seeks a ***story*** [emphasis mine]. Inevitably the story is characterized by a mystery. Since the world does not yield its secrets easily, the scientist must be a careful and persistent observer."

As our tales unfold, discrepant events and unexpected results tickle the characters in the stories and prick their wonder centers, making them ask, "What's going on here?" Most important of all, our stories have endings that are different from most. They are the mysteries that Martin and Miller talk about. They end with a question, an invitation to explore and extend the story and to engage in inquiry.

Our stories do not come with built-in experts who eventually solve the problem and expound on the solution. There is no "Doctor Science" who sets everybody straight in short order. Moms, dads, sisters, brothers, and friends may offer opinionated suggestions ripe for consideration or tests to be designed and carried out. It is the readers who are challenged to become the scientists and solve the problem.

references

American Association for the Advancement of Science (AAAS).1993. *Benchmarks for science literacy.* New York: Oxford University Press.

Bransford, J. D., A. L. Brown, and R. R. Cocking, eds. 1999. *How people learn.* Washington, DC: National Academy Press.

Duckworth, E. 1986. *Inventing density.* Grand Forks, ND: Center for Teaching and Learning, University of North Dakota.

Martin, K., and E. Miller. 1990. Storytelling and science. In *Toward a whole language classroom: Articles from language arts, 1986–1989*, ed. B. Kiefer. Urbana, IL: National Council of Teachers of English.

National Research Council (NRC). 2000. *Inquiry and national science education standards: A guide for teaching and learning.* Washington, DC: National Academy Press.

Osborne, R., and P. Fryberg. 1985. *Learning in science: The implications of children's science.* Auckland, New Zealand: Heinemann.

Research on learning. 1993. *Science Education* 77 (5): 497–541.

Schneps, M. 1996. *The private universe project.* Harvard Smithsonian Center for Astrophysics.

Shapiro, B. 1994. *What children bring to light.* New York: Teachers College Press.

Watson, B., and R. Konicek. 1990. Teaching for conceptual change: Confronting children's experience. *Phi Delta Kappan* 71 (9): 680–684.

CHAPTER 2
USING THE BOOK AND THE STORIES

It is often difficult for overburdened teachers to develop lessons or activities that are compatible with the everyday life experiences of their students. A major premise of this book is that if students can see the real-life implications of science content, they will be motivated to carry out hands-on, minds-on science investigations and personally care about the results. Science educators have, for decades, stressed the importance of science experiences for students that emphasize personal involvement in the learning process. I firmly believe that the use of open-ended stories that challenge students to engage in real experimentation about real science content can be a step toward this goal. Furthermore, I believe that students who see a purpose to their learning and experimentation are more likely to understand the concepts they are studying, and I sincerely hope that the contents of this book will relieve the teacher from the exhausting work of designing inquiry lessons from scratch.

These stories feature children in natural situations at home, on the playground, at parties, in school, or in the outdoors. Children should be able to identify with the story characters, to share their frustrations, concerns and questions. The most important role for the adult is to help guide and facilitate investigations and to debrief activities with them and to think about their analyses of results and conclusions. The children often need help to go to the next level and to develop new questions and find ways of following these questions to a conclusion. Our philosophy of science education is based on our beliefs that children can and want to care enough about problems to make them their own. This should enhance and invigorate any curriculum. In short, students can begin to lead the curriculum, and because of their personal interest in the questions that evolve from their activities, they will maintain interest for much longer than they would if they were following someone else's lead.

Interestingly enough, one of my students, a teacher, says that one of her biggest problems is getting her students to "care" about the topics they are studying. She says they go through the motions but without affect. Perhaps that same problem is not new to you. I hope that this book can help you take a step toward solving that problem. It is difficult if not impossible to make each lesson personally relevant to every student. However, by focusing on everyday situations highlighting kids looking at everyday phenomena, I believe that we can come closer to reaching student interests.

I strongly suggest the use of complementary books as you go about planning for inquiry teaching. Five special books are *Uncovering Student Ideas* (volumes 1, 2, 3, and 4) by page Keeley et al. published by NSTA press and *Science Curriculum Topic Study* by Page Keeley published by Corwin Press and NSTA. The multiple volumes of *Uncovering Student Ideas* help you find out what kinds of preconceptions your students bring to your class. *Science Curriculum Topic Study* focuses on finding the background necessary to plan a successful standards-based unit. I would strongly recommend that you find a copy of *Science Matters: Achieving Scientific Literacy,* by Robert Hazen and James Trefil. This book will become your reference for all matters scientific. It is written in a simple, direct, and accurate matter and will give you the necessary background in the sciences when you need it.

Another especially useful book is *Making Sense of Secondary Science: Research Into Children's Ideas* (Driver et al. 1994). The title of this book is misleading in the term "secondary science." In Great Britain, anything above primary level is referred to as secondary. It is a compilation of the research done on children's thinking about science and is a must have for teachers. In 1978, David Ausubel made one of the most simple but telling comments about teaching: "The most important single factor influencing learning is what the learner already knows; ascertain this, and teach him accordingly." I would also like to alert you to the September 2006 issue of *Science and Children,* which has as its theme teaching investigation skills. In it are articles that will give you many ideas for encouraging and reinforcing these skills.

On page 11 is a chart showing the relationships among several of the various books mentioned above and the stories in this book. Note that each story is matched with the appropriate topic in the *Science Curriculum Topic Study's Curriculum Topic Study Guide* and the probe(s) in the *Uncovering Student Ideas* books, volumes 1–3.

The background material that accompanies each story is designed to help you find out what your learners already know about your chosen topic and what to do with that knowledge as you plan. The above-mentioned books will supplement the materials in this book and deepen your understanding of teaching for inquiry.

How then, is this book set up to help you to plan and teach inquiry based science lessons?

HOW THIS BOOK IS ORGANIZED

The stories are arranged in three sections. There are five stories related to the biological sciences, five for the Earth systems science, and five for the physical sciences. There is a concept matrix at the beginning of each section that can be used to select a story most related to your content need. Following this matrix you will find the stories and the background material in separate chapters. Please note that the Earth systems science stories purposefully integrate the physical and biological sciences into science mysteries that focus on all aspects of everyday science related to the Earth sciences.

Each chapter, starting with Chapter Five, will have the same organizational format. First you will find the story followed by background material for using the story. The background material will contain the following sections:

Story in this book	Curriculum Topic Study Guide	Uncovering Student Ideas in Science		
		Volume 1	Volume 2	Volume 3
What's Hiding in the Woodpile?	Energy Transformation; Energy	Mitten Problem; Objects and Temperature	n/a	Thermometer
The New Greenhouse	Weather and Climate; Solar Energy; Air and Atmosphere	Objects and Temperature	n/a	Thermometer
Rotten Apples	Flow of Energy Through Ecosystems; Cycling of Matter; Fungi and Microorganisms	n/a	Is It Food for Plants?	Earth's Mass; Rotting Apple
Now Just Wait a Minute!	Experimental Design; Variables; Motion; Work, Power, and Machines	n/a	n/a	Is It Scientific Inquiry? Rolling Marbles
Cool It, Dude!	Heat and Temperature; Energy Transformation	Cookie Crumbling? Is It Melting? Ice Cubes in a Bag	Ice Cold Lemonade; Freezing Ice	Ice Cubes in a Glass; Thermometer
Worms Are for More Than Bait	Animal Life; Life Cycles; Populations and Communities; Interdependency Among Organisms; Flow of Energy Through Ecosystems	Is It an Animal? Is It Living?	Is It Food for Plants?	Does It Have a Life Cycle? Rotting Apple
What Did That Owl Eat?	Reproduction, Growth, and Development; Animal Life; Populations and Communities; Food Chains and Food Webs	Is It an Animal? Is It Living?	Habitat Change	n/a
Trees from Helicopters, Continued	Plant Life; Reproduction, Growth, and Development	Is It a Plant?	Needs of Seeds; Is it a Plant?	Where Do Seeds Come From?

Flowers: More Than Just Pretty	Life Processes and Needs of Organisms; Cells; Plant Life	Seeds in a Jar; Functions of Living Things	Is it a Plant?	Does It Have a Live Cycle? Where Do Seeds Come From?
A Tasteful Story	Senses; Human Body Systems; Behavioral Characteristics of Organisms	Human Body Basics; Functions of Living Things	n/a	n/a
The Magnet Derby	Magnetism; Electrical Charge; Energy	Is It Matter?	n/a	n/a
Pasta in a Hurry	Conservation of Energy; Properties of Matter; Chemical Bonding; Heat and Temperature	Going Through a Phase	Boiling Time and Temperature; Turning the Dial; What's in the Bubbles?	Thermometer
Iced Tea	Physical Properties and Change; Mixtures and Solutions	Is It Melting?	Ice Cold Lemonade	Thermometer
Color Thieves	Electromagnetic Spectrum; Visible Light, Color, and Vision; Waves: Color Spectrum	Can It Reflect Light? Apple in the Dark	n/a	n/a
A Mirror Big Enough	Visible Light, Color, and Vision; Waves	Can It Reflect Light?	n/a	Mirror on the Wall

Purpose

This section describes the concepts and/or general topic that the story attempts to address and where it fits into the general scheme of science concepts. It may also place the concepts within a conceptual scheme of a larger idea. For example, in "Rotten Apples" the story is shown to be part of a larger concept of decay, decomposition, and perhaps composting.

Related Concepts

A concept is a word or combination of words that form a mental construct of an idea. Examples are *motion, reflection, rotation, heat transfer,* and *acceleration.* Each story is designed to address a single concept, but often the stories open the door to several concepts. You will find a list of possible related concepts listed in the teacher background material. You should also check the matrices of stories and related concepts.

Don't Be Surprised

In most cases, this section includes projections of what your students will most likely do and how they may respond to the story. The projections relate to the content but focus more on the development of their current understanding of the concept. It may even challenge you to prepare for teaching by doing some of the projected activities yourself, so that you are prepared for what your students will bring to class. For example, with "What's in a Pellet" you may want to dissect an owl pellet yourself before asking your students to do so. In that way you will be prepared for the data they will bring to class and be aware of possible problems.

Content Background

This material a very succinct "short course" on the conceptual material that the story targets. It is not, of course, is a complete coverage but should give you enough information to feel comfortable using the story and getting started on the inquiry. In most instances, references to books, articles, and internet sites are included to help you in preparing yourself to teach the topic. It is important that you have a reasonable knowledge of the topic in order for you to lead the students through their inquiry. It is not necessary, however, for you to be an expert on the topic. Learning along with your students can help you understand how their learning takes place and make you a member of the class team striving for understanding of natural phenomena.

Related Ideas from the National Science Education Standards (NRC) and Benchmarks for Science Literacy (AAAS)

These two documents are considered to be the National Standards upon which most of the local and state standards documents are based. For this reason, the concepts listed for the stories are almost certainly the ones listed in your local curriculum. It is possible that some of the stories are not mentioned specifically in the Standards but are clearly related. I suggest that you obtain a copy of *Science Curriculum Topic Study* (Keeley 2005), which will help you immensely with finding information about content, children's preconceptions, standards, and more resources. Even though it may not be mentioned specifically in each of the stories, you can assume that all of the stories will have connections to the Standards and Benchmarks in the area of Inquiry, Standard A.

Using the Story With Grades K–4 and 5–8

These stories have been tried with children of all ages. We have found that the concepts apply to all grade levels but at different levels of sophistication. Some of the stories have themes and characters that resonate better with one age group than another. However, the stories can be altered slightly to appeal to an older or younger group very simply by changing the characters to a more appropriate age or using a different situation. The theme is the same; just the characters and setting are modified. Please always read the suggestions for both grade levels.

As you may remember from the case study in the introduction, grade level is of little consequence in determining which stories are appropriate at which grade level. Both classes developed investigations appropriate to their developmental abilities. Second graders were satisfied to find out what happens to the length of a tree's shadow over a school year while the middle school class developed more sophisticated investigations involving length of day, direction of shadows over time, and the daily length of shadows over an entire year. The main point here is that by necessity some stories are written with characters more appealing to certain age groups than others. Once again, I encourage you to read both the K–4 and 5–8 sections Using The Story, because the ideas presented for either grade level may be suited to your particular students.

There is no highly technical apparatus to be bought. Readily available materials found in the kitchen, bathroom, or garage will usually suffice. You are provided with background information about the principles and concepts involved and a list of materials you might want to have available. These suggestions of ideas and materials are based upon experience while testing these stories with children. While classrooms, schools, and children differ, most childhood experiences and development result in similar reactions to how you explain and develop questions about the tales. Whether they belong to Jenny, the artistic girl, or Tenika buying a mirror, the problems beg for solutions and, most important, create new questions to be explored by the young scientists.

Here you will find suggestions to help you teach the lessons that will allow your students to become active inquirers, develop their investigations, and finally finish the story, which you may remember was left open for just this purpose. There is no step-by-step approach or set of lesson plans to accomplish this end. Obviously, you know your students, their abilities, their developmental levels, their learning abilities and disabilities better than anyone. You will discover, however, some suggestions and some techniques that were found to work well in teaching for inquiry. You may use them as written or modify them to fit your particular situation. The main point is that you try to involve your students as deeply as possible in trying to solve the mysteries posed by the stories.

Related NSTA Science Store Publications and NSTA Journal Articles

Here, you will find lists of specific books and articles from the constantly growing treasure trove of National Science Teacher Association (NSTA) resources for teachers. While our listings are not completely inclusive, you may access the entire scope of resources on the internet at *www.nsta.org/store*. Membership in NSTA will allow you to read all articles on line.

References

References are provided for the articles and research findings cited in the background section for each story.

Concept Matrices

At the beginning of each section—Earth and space science stories, biological related stories, and physical science related stories—you will find a concept matrix, which indicate the concepts most related to each story. It can be used to select a story that matches your instructional needs.

FINAL WORDS

I was pleased to find that Michael Padilla, past president of NSTA, asked the same questions as I did when I decided to write a book that focused on inquiry. In the May 2006 edition of *NSTA Reports*, Mr. Padilla in his "President's Message" commented, "To be competitive in the future, students must be able to think creatively, solve problems, reason and learn new, complex ideas…[Inquiry] is the ability to think like a scientist, to identify critical questions to study; to carry out complicated procedures, to eliminate all possibilities except the one under study; to discuss, share and argue with colleagues; and to adjust what you know based on that social interaction." Further, he asks, "Who asks the question?...Who designs the procedures?...Who decides which data to collect?...Who formulates explanations based upon the data?...Who communicates and justifies the results?...What kind of classroom climate allows students to wrestle with the difficult questions posed during a good inquiry?"

I believe that this book speaks to these questions and that the techniques proposed here are one way to answer the above questions with "The students do!" in the kind of science classroom this book envisions.

REFERENCES

Ausubel, D., J. Novak, and H. Hanensian. 1978. *Educational psychology: A cognitive view.* New York: Holt, Rinehart, and Winston.

Driver, R., A. Squires, P. Rushworth, and V. Wood-Robinson. 1994. *Making sense of secondary science: Research into children's ideas.* London and New York: Routledge Falmer.

Hazen, R., and J. Trefil. 1991. *Science matters: Achieving scientific literacy.* New York: Anchor Books.

Keeley, P. 2005. *Science curriculum topic study: Bridging the gap between standards and practice.* Thousand Oaks, CA: Corwin Press.

Keeley, P., F. Eberle, and L. Farrin. 2005. *Uncovering student ideas in science: 25 formative assessment probes.* Arlington, VA: NSTA Press.

Keeley, P., F. Eberle, and J. Tugel, J. 2007. *Uncovering student ideas in science: 25 more formative assessment probes.* Arlington, VA: NSTA Press.

Padilla, M. *NSTA Reports* President's Message. Teaching Investigation Skills. 2006. *Science and Children* 44(1).

USING THE BOOK IN DIFFERENT WAYS

Although the book was originally designed for use with K-8 students by teachers or adults in informal settings, it became obvious that a book containing stories and content material for teachers intent on teaching in an inquiry mode had other potential uses. I list a few of them below to show that the book has several uses beyond the typical elementary and middle school populations in formal settings.

USING THE BOOK AS A CONTENT CURRICULUM GUIDE

When asked by the University of Massachusetts to teach a content course for a special master's degree program in teacher education, I decided to use *Everyday Science Mysteries* as one of several texts to teach content material. A major premise in the book is that students, when engaged in answering their own questions, will delve into a topic at a level commensurate with their intellectual development and learning skills. Therefore, even though the stories were designed for people younger than themselves, the students in the class were able to find questions to answer that were at a level of sophistication that challenged them.

During the fall 2007 semester this book was used as a text and curriculum guide for a class titled Exploring the Natural Sciences Through Inquiry at the University of Massachusetts in Amherst. The shortened version of the syllabus for the course follows:

Exploring the Natural Sciences Through Inquiry
EDUC 692 O
Fall 2007

Instructor: Dr. Richard D. Konicek, Professor Emeritus
Time: Thursdays, 4–7 p.m.
Location: High School of Science and Technology, Springfield, MA

Course Description:

This course is designed for elementary and middle school teachers who need, not only to deepen their content knowledge in the natural sciences, but also to understand how inquiry can be used in the elementary and middle school classroom. Natural sciences mean the Biological Sciences, Earth and Space Sciences, and the Physical Sciences. Teachers will sample various topics from each of the above areas of science through inquiry techniques. The topics will be chosen from everyday phenomena such as Astronomy (Moon and Sun observations) Physics (motion, energy, thermodynamics, sound, periodic motion,)and Biology (botany, zoology, animal and plant behavior, evolution).

Course Objectives:

It is expected that each student will

- gain content background in each of the three areas of natural science;
- be able to apply this content to their teaching methods;
- develop questions concerning a particular phenomenon in nature;
- design and carry out experiments to answer their questions;
- analyze experimental data and draw conclusions;
- consult various sources to verify the nature of their conclusions;
- read scientific literature appropriate to their studies; and
- extend their knowledge to use with middle school children both in content and methodology.

Relationship to the Conceptual Framework of the School of Education:

Collaboration: Teachers will work in collaborative teams during class meetings to acquire science content and pedagogical knowledge and skills.

Reflective Practice: Teachers will develop and implement formative assessment probes with their students.

Multiple Ways of Knowing: Teachers will share science questions and their methods of inquiry chosen to answer those questions.

Access, Equity, & Fairness: Teachers will reflect on student understandings based on students' stories.

Evidence-Based Practice: Teachers will explore formative assessment through the use of probes.

Required Texts:

Konicek-Moran, R. 2007. *Everyday science mysteries.* Arlington, VA: NSTA Press.

Hazen, R. M., and J. Trefil. 1991. *Science matters.* New York: Anchor Books.

Keeley, P., F. Eberle, and J. Tugel. 2007. *Uncovering student ideas in science: 25 more formative assessment probes* (vol. 2). Arlington, VA: NSTA Press.

Resource Texts:

American Association for the Advancement of Science (AAAS). 2001. *Atlas of science literacy* (vol. 1). Washington, DC: AAAS.

American Association for the Advancement of Science (AAAS). 2007. *Atlas of science literacy* (vol. 2). Washington, DC: AAAS.

Driver, R., A. Squires, P. Rushworth, and V. Wood-Robinson. 1994. *Making sense of secondary science.* London: Routledge-Falmer.

Keeley, P., F. Eberle, and L. Farrin. 2005. *Uncovering student ideas in science.* Arlington VA: NSTA Press.

Topics to Be Investigated in Volume One:

Everyday Science Mysteries is organized around stories. The core concepts related to the National Science Education Standards developed by the National Research Council in 1996 are the basis for the concept selection. The story titles and related core concepts are shown in the matrices below.

	Earth Systems Science				
	Stories				
Core Concepts	Moon Tricks	Where Are the Acorns?	Master Gardener	Frosty Morning	The Little Tent That Cried
States of Matter			X	X	X
Change of State			X	X	X
Physical Change			X	X	X
Melting			X	X	
Systems	X	X	X	X	X
Light	X	X			
Reflection	X	X		X	
Heat Energy			X	X	X
Temperature				X	X
Energy			X	X	X
Water Cycle				X	X
Rock Cycle			X		
Evaporation				X	X
Condensation				X	X
Weathering			X		
Erosion			X		
Deposition			X		
Rotation/Revolution	X	X			
Moon Phases	X				
Time	X	X			

| Physical Sciences | | | | | |
| | | Stories | | | |
Core Concepts	Magic Balloon	Bocce Anyone?	Grandfather's Clock	Neighborhood Telephone Service	How Cold Is Cold?
Energy	X	X	X	X	X
Energy Transfer	X	X	X	X	X
Conservation of Energy		X			X
Forces	X	X	X		
Gravity	X	X	X		
Heat	X				X
Kinetic Energy		X	X		
Potential Energy		X	X		
Position and Motion		X	X		
Sound				X	
Periodic Motion			X	X	
Waves				X	
Temperature	X				X
Gas Laws	X				
Buoyancy	X				
Friction		X	X		
Experimental Design	X	X	X	X	X
Work		X	X		
Change of State					X
Time		X	X		

National Science Teachers Association

Core Concepts	Biological Sciences				
	Stories				
	About Me	Bugs	Dried Apples	Seed Bargains	Trees From Helicopters
Animals	X	X			
Classification		X	X	X	X
Life Processes	X	X	X	X	X
Living Things	X	X	X	X	X
Structure and Function		X	X		X
Plants			X	X	X
Adaptation		X			X
Genetics/Inheritance	X		X	X	X
Variation	X		X	X	X
Evaporation			X		
Energy		X	X	X	X
Systems	X	X	X		X
Cycles	X	X	X	X	X
Reproduction	X	X	X	X	X
Inheritance	X	X	X		X
Change		X	X		
Genes	X		X		X
Metamorphosis		X			
Life Cycles		X	X		X
Continuity of Life	X	X	X	X	X

Assignments:

Astronomy (25%): Everyone will be expected to explore the daytime astronomy sequence, which will aim to develop models of the Earth, Moon, and Sun relationships. Students will keep a Moon journal and Sun shadow journal over the course of the semester, which they will turn in periodically.

Topics (50%): In addition, students will pick at least two topics from each of the Earth, Physical, and Biological areas for study during the semester. Students will come up with a topic question and do an investigation or experiment regarding the topic questions posed. (For example: Are there acorns that do not need a dormancy period before germinating?) These questions and experiments will be shared with the class as they progress so that all students will either be directly involved in learning about the content or indirectly involved by listening to reports and critiquing those reports. In addition to the experiments, students will (1) involve their students in their experiments/investigations and (2) design and give formative assessment probes to their students to find out what knowledge they already possess. Students will be graded on their experimental designs, their presentations of their data, their conclusions. I will develop a rubric with the students that will address the goals stated above and their values to be calculated for their grades. Attendance/Participation (25%): Attendance at all course meetings is required.

References for Course Development:

American Association for the Advancement of Science (AAAS).1993. *Benchmarks for science literacy*. New York: Oxford University Press.

Ausubel, D., J. Novak, and H. Hanensian. 1978. *Educational psychology: A cognitive view.* New York: Holt, Rinehart and Winston.

Bransford, J. D., A. L. Brown, and R. R. Cocking, eds. 1999. *How people learn*. Washington, D.C. National Academy Press.

Driver, R., A. Squires, P. Rushworth, and V. Wood-Robinson. 1994. *Making sense of secondary science: Research into children's ideas.* London and New York: Routledge Falmer.

Duckworth, E. 1986. *Inventing density*. Grand Forks, ND: Center for Teaching and Learning, University of North Dakota.

Hazen, R., and J. Trefil. 1991. *Science matters: Achieving scientific literacy.* New York: Anchor Books.

Keeley, P. 2005. *Science curriculum topic study: Bridging the gap between standards and practice.* Thousand Oaks, CA: Corwin Press.

Keeley, P., F. Eberle, and L. Farrin. 2005. *Uncovering student ideas in science: 25 formative assessment probes* (vol. 1). Arlington, VA: NSTA Press.

Keeley, P., F. Eberle, and J. Tugel. 2007. *Uncovering student ideas in science: 25 more formative assessment probes* (vol. 2). Arlington, VA: NSTA Press.

Konicek-Moran, R. 2008. *Everyday science mysteries*. Arlington, VA: NSTA Press.

Martin, K., and E. Miller. 1990. Storytelling and science. In *Toward a whole language classroom: Articles from Language Arts,* ed. B. Kiefer. Urbana, IL: National Council of Teachers of English.

National Research Council (NRC). 2000. *Inquiry and national science education standards: A guide for teaching and learning*. Washington, DC: National Academy Press.

Osborne, R., and P. Fryberg. 1985. *Learning in science: The implications of children's science.* Auckland, New Zealand: Heinemann.

Schneps, M. 1996. *The private universe project*. Washington, DC: Harvard Smithsonian Center for Astrophysics.

Shapiro, B. 1994. *What children bring to light*. New York: Teachers College Press.

Watson, B., and R. Konicek. 1990. Teaching for conceptual change: Confronting children's experience. *Phi Delta Kappan* May: 680–684.

The course was taught as a graduate course for teachers or prospective teachers of elementary or middle students. The course could be classified as a content/pedagogy class for teachers who had minimal science backgrounds as well as minimal skills in teaching for inquiry. My premise was that if teachers would learn content through inquiry techniques, they would be convinced of their efficacy as learning techniques and would be likely to use them to teach content in their own classes. As it turned out, those teacher-students who had classes of their own and were full-time teachers did work on their projects with their students with very satisfactory results, according to the teachers. As a result, both teachers and students were learning science content through inquiry techniques. Because the teachers in the class were completing an assignment, they were able to be honest with their students about not knowing the outcome of their investigations. This is often a problem with teachers who are afraid to admit that they are learning along with the students. In this case, the students were excited about learning along with their teachers and vice versa. Teachers with classrooms were also able to develop rubrics with their students for the grading of their explorations and therefore were involved with some metacognition as well.

As a result of this small foray, I am convinced that this book can be used as a content guide for undergraduate and graduate content-oriented courses for teachers. As noted in the sample syllabus, the use of other supplementary texts for content and pedagogy adds to the strength of the course in preparing teachers to use inquiry techniques and to learn content themselves. With the use of the internet, very little information is hidden from anyone with minimum computer skills. Unlike many survey courses chosen by teachers who are science-phobic, this course did not attempt to cover a great number of topics but to teach a few topics for understanding. The basic premise of this author is that when deciding between coverage and understanding science topics and concepts, understanding wins every time. It is well known that our current curriculum in the United States has been faulted for being a mile wide and an inch deep. High stakes testing seems to also add to the problem since almost all teachers whom I have interviewed over the past few years are reluctant to teach for understanding using inquiry methods because teaching for understanding takes more time and does not allow for coverage of the almost infinite amount of material that might appear on a standardized tests. Thus, student misconceptions are seldom addressed and continue to persist even though students can do reasonably well on teacher-made tests and assessment tools and still hold on to their misconceptions. See Bonnie Shapiro's book, *What Children Bring to Light.*

USING THIS BOOK as a resource FOR science METHODS COURSES

Traditionally, science methods courses in the United States are taught to classes mainly composed of science-phobic students. One of the main goals of science methods courses is to make students comfortable with science teaching and to help students develop skills in teaching science to youngsters using a hands-on, minds-on approach. Unfortunately, a great many students come to these methods courses with a minimum of science content courses, and many of those are either survey (non-laboratory) courses or courses taught in large-lecture format. In 12–13 weeks, methods instructors are expected to convert these students into confident, motivated teachers who are familiar with techniques that promote inquiry learning among their students. Having taught this type of course to undergraduates and career-changing graduate students for more than 30 years, I have found that making students comfortable with science is the first goal. This is often accomplished by assigning students science tasks that can be accomplished with a minimum of stress and with a maximum of success. Second, I try to instill the ideas commensurate with the nature of science as a discipline. Third, I find that it is often necessary to teach a little content for those who are rusty and to clarify some misconceptions. Last, but not least important, I try to acquaint them with resources in the field so that they know what is available to them as they enter their teaching careers. Obviously, here is an opportunity to acquaint them with current information about the learners themselves, how students learn, and how best to teach for inquiry.

As a final assignment in my methods classes, I assign the students the task of writing an everyday science mystery and a paper to accompany it that will describe, how they will use the story to teach a concept using the inquiry approach. The results have far exceeded what I had been receiving from the typical lesson plan used by others and me through the years. This book would not only provide the text on teaching science (found in the early chapters) but would provide a model for producing everyday science mysteries for topics of the students' choices.

USING THIS BOOK IN HOMESCHOOL PROGRAMS

Homeschooling parents have a great many resources at their disposal, as any internet search will show. Curricular suggestions and materials are available for those parents and children who choose to conduct their education at home. Science is one of those subjects that might be difficult for many parents whose science backgrounds are a bit weak or outdated. Parents and children working together to solve a story-driven mystery could use this book easily. The connections to the National Science Education Standards also help in making sure that the homeschooling curriculum is covering the nationally approved scientific concepts. Parents would use the book just as any teacher would use it, except there would be fewer opportunities for class discussions and the parents would have to do a bit more discussion with their children to solidify their understanding of their investigations.

references

Keeley, P., F. Eberle, and C. Dorsey. 2008. *Uncovering student ideas in science: Another 25 formative assessment probes.* Arlington, VA: NSTA Press.

Klentschy, M. 2008. *Using science notebooks in elementary classrooms.* Arlington, VA: NSTA Press.

Leach, J. T., R. D. Konicek, and B. L. Shapiro. 1992. The ideas used by British and North American School children to interpret the phenomenon of decay: A cross-cultural study. Paper presented to the Annual Meeting of the American Educational Research Association. San Francisco.

Osborne, R., and P. Fryberg. 1985. *Learning in science: The implications of children's science.* Auckland, New Zealand: Heinemann.

Shapiro, B. 1994. *What children bring to light.* New York: Teachers College Press.

Schneps, M. 1996. *The private universe project.* Washington, DC: Harvard Smithsonian Center for Astrophysics.

THE LINK BETWEEN SCIENCE, INQUIRY, AND LANGUAGE LITERACY

While heading into the final chapter before launching into the stories, I couldn't resist introducing you to a piece of literature that is seldom read except by English literature majors. The quotation that follows is from Irish novelist James Joyce in his classic book, *Ulysses*, written in 1922:

Where was the chap I saw in that picture somewhere? Ah, in the dead sea, floating on his back, reading a book with a parasol open. Couldn't sink if you tried: so thick with salt. Because the weight of the water, no, the weight of the body in the water is equal to the weight of the. Or is it the volume is equal to the weight? It's a law something like that. Vance in High school cracking his fingerjoints, teaching. The college curriculum. Cracking curriculum. What is weight really when you say weight? Thirtytwo feet per second, per second. Law of falling bodies: per second, per second. They all fall to the ground. The earth. It's the force of gravity of the earth is the weight. (p.73)

In his novel, Joyce's main character Bloom recalls a picture of someone floating in the Dead Sea, and tries to recall the science behind his fascination with the event. Have you or have you observed others who, while trying to explain something scientific, resorted to this recall of a mishmash of scientific knowledge, half-remembered and garbled? (For this foray into literature, I am indebted to Michael J. Reiss who called my attention to this passage while recently reading an article of his in *School Science Review*.)

In his school days, Bloom seems to have been fascinated both with the curriculum and the teacher in his physics class. However, Bloom's memory of the science behind buoyancy runs the gamut from unrelated science language pouring out of his memory bank to visions of his teacher cracking his finger joints. Unfortunately, even today, this might well be the norm rather than the exception. This phenomenon is exactly what we are trying to avoid in our modern pedagogy and now leads us into the main point of this chapter.

There are many ways of connecting literacy and science. We shall look briefly at the research literature and find some ideas that will make the combination of literacy and science not only worthwhile but also essential for learning.

LiTeracy anD science

In pedagogical terms there are differences between science literacy and the curricular combination of science and literacy, but perhaps they have more in common than one might expect. Science literacy is the ability to understand scientific concepts so that they have a personal meaning in everyday life. In other words, a science literate population can use their knowledge of scientific principles in situations other than those in which they learned them. For example, I would consider people science literate if they were able to use their understanding of ecosystems and ecology to make informed decisions about saving wetlands in their community. This is of course what we would hope for in every aspect of our educational goals regardless of the subject matter. Literacy refers to the ability to read, write, speak, and make sense of text. Since most schools emphasize reading and writing and mathematics, they often take priority over all other subjects in the school curriculum. How often have I heard teachers say that their major responsibility is reading and math and that there is no time for science? But there is no need for competition for the school day. I believe that this form of competition is caused by the lack of understanding of the synergy created by integration of subjects. In synergy, you get a combination of skills that surpasses the sum of the individual parts.

So what does all of this have to do with teaching science as inquiry? There is currently a strong effort to combine science and literacy, because a growing body of research stresses the importance of language in learning science. Recall, if you will, that hands-on science is nothing without its minds-on counterpart, since a food fight is a hands-on activity but one does not learn much through mere participation, except perhaps the finer points of the aerodynamic properties of Jell-O. The understanding of scientific principles is not embedded in the materials themselves or in the manipulation of these materials. Discussion, argumentation, discourse of all kinds, group consensus, and social interactions—all forms of communication are necessary for students to make meaning out of the activities in which they have engaged. And these require language in the form of writing, reading, and particularly speaking. They require that students think about their thinking, that they hear their own and others' thoughts and ideas spoken out loud, and that perhaps eventually they see them in writing to make sense of what

they have been doing and the results they have been getting in their activities. This is the often forgotten "minds-on" part of the "hands-on, minds-on" couplet. Consider the following:

> In schools, talk is sometimes valued and sometimes avoided, but—and this is surprising—talk is rarely taught. It is rare to hear teachers discuss their efforts to teach students to talk well. Yet talk, like reading and writing, is a major motor—I could even say the major motor—of intellectual development. (Calkins 2000, p.226)

For a detailed and very useful discussion of talk in the science classroom, I refer you to Jeffrey Winokur and Karen Worth's chapter, "Talk in the Science Classroom: Looking at What Students and Teachers Need to Know and Be Able to Do," in *Linking Science and Literacy in the K–8 Classroom* (2006).

The concept of linking inquiry-based science and literacy has a strong intellectual and research base. First, the theoretical work of Padilla and his colleagues suggests that inquiry science and reading share a set of intellectual processes (e.g., observing, classifying, inferring, predicting, and communicating) and that these processes are used whether the student is conducting scientific experiments or reading text (Padilla, Muth, and Padilla 1991). Helping children become aware of their thinking as they read and investigate materials will help them understand and practice more metacognition. You may have to model this for them by thinking out loud yourself as you view a phenomenon. Help them understand why you spoke as you did and why it is important to think about your process of thinking. You may say something like, "I think that warm weather affects how fast seeds germinate. I think that I should design an investigation to see if I am right." Then later, "Did you notice how I made a prediction/hypothesis that I could test in an experiment?" Modeling your thinking can help your students see how and why the talk of science is used in certain situations. Science is about words and their meanings.

Postman made a very interesting statement about words and science. He said, "Biology is not plants and animals. It is language about plants and animals…. Astronomy is not planets and stars. It is a way of talking about planets and stars" (1979, p.165). To emphasize this point even further, I might add that science is a language, a language that specializes in talking about the world and being in that world we call science. It has a special vocabulary and organization and scientists use this vocabulary and organization when they talk about their work. Often, it is called "discourse" (Gee 2004). Children need to learn this discourse when they present their evidence, when they argue the fine points of their work, evaluate their own and others works, and refine their ideas for further study. Students do not come to you with this language in full bloom; in fact the seeds may not even have germinated. They attain it by doing science and being helped by knowledgeable adults who teach them about controlling variables or fair tests, about having evidence to back up their statements, and about using the processes of science in their attempts at what has been called "first hand inquiry" (Palincsar and Magnusson, 2001).

This is inquiry that involves the direct involvement with materials or in other more familiar words, the hands-on part of scientific investigation. The term "secondhand investigations" refers to the use of textual, lecture, reading data, charts, graphs, or other types of instruction that do not feature direct contact with materials. Cervetti et al. put it so well:

> [W]e view firsthand investigations as the glue that binds together all of the linguistic activity around inquiry. The mantra we have developed for ourselves in helping students acquire conceptual knowledge and the discourse in which that knowledge is expressed (including particular vocabulary) is "read it, write it, talk it, do it!"—and in no particular order, or better yet, in every possible order. (2006, p. 238)

So you can see that it is also important that the students talk about their work, write about their work, read about what others have to say about the work they are doing (in books or via visual media), and take all possible opportunities to document their work in a way that is useful to them in looking back at what they have found out about their work.

SCIENCE NOTEBOOKS

Many science educators have lately touted science notebooks as an aid to involving students more in the discourse of science (Klentschy 2005; Klentschy 2008; Campbell and Fulton 2003). Their use has also shown promise in helping English language learners (ELLs) in the development of language skills as well as in the learning of science concepts and the nature of science (Klentschy 2005).

Science notebooks differ from science journals and science logs because they are not merely for recording data (logs) or for demonstrating learning (journals) but are meant to be used continuously for recording investigations, designs, plans, thinking, vocabulary, and concerns or puzzlement. The science notebook is the recording of past, present, and future thoughts and is unique to each student. The teacher makes sure that the students have ample time to record events and to also ask for specific responses to such questions as "What still puzzles you about this activity?"

For specific ideas for using science notebooks and for information of the value of using the notebooks in science, see *Science Notebooks: Writing About Inquiry* by Brian Campbell and Lori Fulton (2003) and "Science Notebook Essentials: A Guide to Effective Notebook Components," an article in *Science and Children*, by Michael Klentschy (2005). In each chapter of this book, the teacher background material will refer to uses of the notebooks. Michael Klentschy has also written a book titled *Using Science Notebooks in Elementary Classrooms* in which he gives many practical examples of student work in science notebooks (Klentschy 2008).

I must acknowledge here the experience I gained in working with a teacher years ago in Pelham, Massachusetts, Dr. Marna Bunce-Crim. It was in her classroom that I learned the power of writing across the curriculum. In her classroom I witnessed minor miracles of children writing to learn, and I came away with a

great appreciation for the power of literacy in science education, especially the importance of asking children to write each day about something that still confused them. The results were remarkable, and as we read their notebooks we witnessed their metacognition, their solutions through their thinking "out loud" in their writing, and, in many cases, the solving of their confusions right there on paper. The use of science notebooks should be an opportunity for the students to record their cognitive journey through their activities. In the case of their use with the stories in this book, it would include the specific question that the student is concerned with, the lists of ideas and statements generated by the class after the story is read, pictures or graphs of data collected by the student and by the class, and perhaps the final conclusions reached by the student and the class as they try to solve the mystery.

Let us imagine that your class has agreed on a conclusion for the story they have been using and that they have reached consensus on that conclusion. What options are open to you as a teacher for asking the students to finalize their work? It may be acceptable to have the students actually write the "ending" to the story or, alternatively, write up the conclusions in a standard lab report format. The former method, of course, is another way of actually connecting literacy and science. Many teachers prefer to have their students at least learn to write the "boiler plate" lab reports, just to be familiar with that method, while others are comfortable with having their students write more anecdotal kinds of reports. My experience is that when students write their conclusions in anecdotal form while referring to their data to support their conclusions, there is more convincing evidence that they have really understood the conceptual understanding they have been chasing rather than filling in the blanks in a form. In the end, it is up to the classroom teacher to decide. Of course, it could be done both ways and two goals might be achieved. While reading the results to Page Keeley's probes given to students, I have become more and more aware that even though the correct answer might or might not have been given in the objective portion of the probe, the true understanding of the concepts are most often visible in the written portion of the probe. (Keeley et al. 2005; Keeley et al. 2007)

As mentioned earlier, a major factor in designing these stories and follow-up activities is based upon one of the major tenets of a philosophy called constructivism. That major tenet is that knowledge is constructed by individuals in order to make sense out of the world in which they live. If we believe this, then the knowledge that each individual brings to any situation or problem must be factored into the way that person tries to solve that problem. By the same token, it is important to realize that the identification of the problem and the way the problem is viewed are also factors determined by each individual. Therefore it is vital that the teacher encourages the students to bring into the open (orally and in writing) those ideas they already have about the situation being discussed. In bringing these preconceptions out of hiding, so to speak, all of the children and the teacher can address the alternative ideas about topics and analyze data openly without hidden agendas in children's minds.

The stories also point out that science is a social, cultural, and therefore human enterprise. The characters in our stories usually enlist others in their investigations, their discussions, and their questions. These people have opinions and hypotheses and are consulted, involved, or drawn into an active dialectic. Group work is encouraged, which in a classroom would suggest cooperative learning. At home, siblings and parents may become involved in the activities and engage in the dialectic as a family group.

The stories can also be read to the children. In this way children can gain more from the literature than if they had to read it by themselves. Children's listening vocabulary is usually greater than their reading vocabulary. Unfamiliar vocabulary can be deduced by the context in which it is found. New vocabulary words can be explained as the story is read. Teachers have found that children are always ready to discuss the stories during the reading and therefore become more involved as they take part in the reading. So much the better because getting involved is what this book is all about—getting involved in situations that beg for problem finding, problem solving, and construction of new ideas about science in everyday life.

WORKING WITH ENGLISH LEARNER POPULATIONS

Suppose that part of your class is made up of students from other cultures and with limited knowledge of the English language. Of what use is inquiry science to such students and how can you use the discipline to increase both their language learning and their science skills and knowledge?

First of all, let's take a look at the problems associated with learning while having limited language understanding. Lee (2005), in his summary of research on ELL students and science learning, points to the fact that students who are not from the dominant society (e.g. Western science) are not aware of the rules and norms of that dominant society. They may come from cultures in which questioning (especially of elders) is not encouraged and where inquiry is not encouraged. Obviously, to help these children cross over from the culture of home to the culture of school, the rules and norms of the new culture must be explained carefully and visibly and the students must be helped to take responsibility for their own learning. There are a number of books written about this topic, and I would not be able to cover the problem in this chapter as well as they have. You can find specific help in a recent NSTA publication titled *Science for English Language Learners: K–12 Classroom Strategies* (Fathman and Crowther 2006). Also very helpful is another NSTA publication, *Linking Science and Literacy in the K–8 Classroom* (Douglas et al. 2006), specifically chapter 12, "English Language Development and the Science-Literacy Connection." Finally, an article from *Science and Children* titled "Teaching Science to English-as-Second Language Learners" (Buck 2000) has many useful suggestions for working with ELL students.

I'll summarize a few ideas as best as I can and will also put them into the teacher background when appropriate:

- Experts agree that vocabulary building is very important for ELL students. You can focus on helping ELL students identify objects they will be working with in their native language and in English. These words can be entered in science notebooks and some teachers have been successful in

using a teaching device called a "working word wall." This is an ongoing poster with graphics and words that are added to the poster as the unit progresses. When possible, real items are taped to the poster. This is visible for constant review, since if it is kept in a prominent location it is helpful for all students, not just the ELL students.

- Many teachers suggest that the group work afforded by inquiry teaching helps ELL students understand the process and the content. Pairing ELL students with English speaking students will facilitate learning since students are often more comfortable receiving help from peers than from the teacher. They are more likely to ask questions of peers than of the teacher because of being more comfortable with peers. It is also likely that explanations from peers may be more helpful because their fellow students explain things in language more suitable to those of their own age and development.

- Use the chalkboard or whiteboard more often. Connect visuals with vocabulary words. Remember that science depends upon the language of discourse. You might also consider inviting parents into the classroom so that they can witness what you are doing to help their children learn English and science. Spend more time focusing on the process of inquiry so that the ELL students will begin to understand how they can take control over their own learning and problem solving. All students can benefit from being considered Science Language Learners (SLLs).

HELPING YOUR STUDENTS AS SLLS

How much help should you give to your students as they work through the problem? A good rule of thumb is that you can help them as much as you think necessary as long as the children are being challenged. In other words, the children should not be following your lead but their own leads. If some of these leads end up in dead ends, then a large part of scientific investigations is part of their experience too. Science is full of experiences that are not productive. If children read most popular accounts of scientific discovery, they would get the impression that the scientist gets up in the morning, asks, "What will I discover today?" and then sets off on a clear, straight path to an elegant conclusion before supper time rolls around. Nothing could be further from the truth. But, it is very important to note that a steady diet of frustration can squash a budding love of science. Dead ends can usually be looked upon as signaling a need for a new design or to ask the question in a different way. Most important, dead ends should not be looked upon as failures. They are more like opportunities to try again in a different way with a clean slate. The adult's role is to keep a balance so that motivation is maintained and interest continues to flourish. Sometimes this is more easily accomplished when kids work in groups. Scientists, too, work in teams and use one another's expertise in a group process.

Many people do not understand that the scientific process includes luck, personal idiosyncrasies, and feelings, as well as the so-called scientific method. The most important aid you can provide your students is to help them maintain their confidence in their abilities to solve problems. They can use metaphors, visualizations, drawings, or any method with which they are comfortable to develop new insights into the problem. Then they can set up their study in a way that reflects the scientific paradigm, including creating a simple question, controlling variables, and isolating the one variable they are testing.

You can also help them keep their experimental designs simple and carefully controlled, and learn to keep good data records in their science notebooks. Most students don't readily see the need for this last point, even after they have been told. They don't see the need because the neophyte experimenter has not had much experience with collecting usable data. Until they realize that unreadable data or necessary data not recorded can cause problems, they see little use for them. Children don't see the need for keeping good shadow length records because they are not always sure what they are going to do with them in a week or a month from now. If they are helped to see the reasons for collecting data and that these data are going to be evidence of a change over time, then they will see the purpose of being able to go back and revisit the past in order to compare it to the present. In this way they can also see the reasons for keeping a log in the first place.

In experiences I have had with children, forcing them to use prescribed data collection worksheets has not helped them understand the reasons for data collection at all and in some cases has actually caused more confusion or amounted to little more than busywork. On one occasion while circulating in a classroom where children were engaged in a worksheet-directed activity, an observer asked a student what she was doing. The student replied without hesitation, "Step Three." Our goal is to empower students engaged in inquiry to the point where they are involved in the activity at a level where all of the steps (including Step Three), are designed by the students themselves and for good reason—to answer their own questions in a logical, sequential, and meaningful manner. We believe it can be done, but it requires patience on the part of the adult facilitators and faith that the children have the skills to carry out such mental gymnastics, with a little help from their friends and mentors.

One last word about data collection: Over the years of being a scientist and working with scientists, one common element stands out. Scientists keep on their person a notebook, which is retrieved numerous times during the day to record interesting items. Memory is seen as an ephemeral thing, not to be trusted. Scientists' notebooks are treasured and essential parts of the scientific enterprise. They don't leave home without them.

And now, on to the stories, which I hope will inspire your students to become active inquirers and enjoy science as an everyday activity in their lives.

references

Buck, G. 2000. Teaching science to English-as-second language learners. *Science and Children* 38 (3): 38–41.

Calkins, L. 2000. *The Art of teaching reading.* Boston: Allyn and Bacon.

Campbell, B., and L. Fulton. 2003. *Science notebooks: Writing about inquiry.* Portsmouth, NH: Heinemann.

Douglas, R., M. P. Klentscly, and K. Worth, eds. 2006. *Linking science and literacy in the K–8 classroom.* Arlington, VA: NSTA Press.

Cervetti, G., P. Pearson, M. Bravo, and J. Barber. 2006. Reading and writing in the service of inquiry-based science. In *Linking science and literacy in the K–8 classroom,* eds. R. Douglas, M. Klentschy, and K. Worth, 221–244, Arlington, VA: NSTA Press.

Fathman, A., and D. Crowther. 2006. *Science for English language learners: K–12 classroom strategies.* Arlington, VA: NSTA Press.

Gee, J. 2004. Language in the science classroom: Academic social languages as the heart of school-based literacy. In *Crossing borders in literacy and science instruction: Perspectives on theory and practice,* ed. E. W. Saul, 13–32. Neward, DE: International Reading Association.

Joyce, J. 1922. *Ulysses.* Repr., New York: Vintage, 1990. New York: Vintage. Page reference is to the 1990 edition.

Keeley, P., F. Eberle, and L. Farrin. 2005. *Uncovering student ideas in science: 25 formative assessment probes* (vol. 1). Arlington, VA: NSTA Press.

Keeley, P., F. Eberle, and J. Tugel. 2007. *Uncovering student ideas in science: 25 more formative assessment probes* (vol. 2). Arlington, VA: NSTA Press.

Klentschy, M. 2005. Science notebook essentials: A guide to effective notebook components. *Science and Children* 43 (3): 24–27.

Klentschy, M 2008. *Using science notebooks in elementary classrooms.* Arlington, VA: NSTA Press.

Padilla M., K. Muth, and R. Padilla. 1991. Science and reading: Many process skills in common? In *Science learning: Processes and applications*, eds. C. M. Santa and D. E. Alvermann, 14–19. Newark, DE: International Reading Association.

Palincsar, A., and S. Magnusson. 2001. The interplay of firsthand and text-based investigations to model and support the development of scientific knowledge and reasoning. In *Cognition and instruction: Twenty-five years of progress*, eds. S. Carver and D. Klahr, 151–194. Mahwah, NJ: Lawrence Erlbaum.

Postman, N. 1979. *Teaching as a conserving activity.* New York: Delacorte.

Reiss, M. 2002. Reforming school science education in the light of pupil views and the boundaries of science. *School Science Review* 84 (307).

Winnokur, J., and Worth, K. 2006. Talk in the science classroom: Looking at what students and teachers need to know and be able to do. In *Linking Science and Literacy in the K-8 Classroom,* eds. R. Douglas, M. Klentschy, and K. Worth, 43–58. Arlington, VA: NSTA Press.

THE
STORIES
AND
BACKGROUND
MATERIAL
FOR
TEACHERS

EARTH SYSTEMS SCIENCE AND TECHNOLOGY

Earth Systems Science and Technology

Core Concepts	What's Hiding in the Woodpile?	The New Greenhouse	Rotting Apples	Now Just Wait a Minute!	Cool It, Dude!
States of Matter					X
Phase Change					X
Heat Energy	X	X			X
Physical Change					X
Energy Spectrum		X			
Temperature		X			X
Conservation of Matter		X	X		X
Living Things			X		
Nature of Technology				X	
Design and Systems				X	
Gravity				X	
Time				X	
Technological Design				X	
Flow of Energy	X				X
Recycling Matter			X		
Weather and Climate		X			
Light		X			

CHAPTER 5

WHAT'S HIDING IN THE WOODPILE?

The *Rug Rats* cartoon was just getting started as Maddie and Justin were settling in for a cozy morning. The room was warm and bright. The wood stove gave off a heat you could almost see. They loved sitting on cushions here and watching TV on Saturday mornings.

Suddenly they were interrupted by Mom, who came in to put more wood in the stove.

"Uh oh," she exclaimed! "Somebody's been forgetting their chores. The wood box is empty and it needs to be filled right away. That is if folks want to enjoy the nice warm room."

The children pretended to be very involved in the show. No one spoke.

"Ahem," said Mom. "Ahem!" she said a little louder. "Wood? We need wood, now!"

Justin looked up. He sat right next to the huge empty wood box and knew what was going to happen. They were going to have to trudge outside into the frosty morning, wheel loads of wood into the garage, and carry it from the garage into the family room. Then it needed to be stacked in the wood box. From the look on Mom's face, it was going to happen soon.

"You know, if you carried a little bit in each day," Mom explained for about the hundredth time, "it wouldn't be such a big chore. But you let it get empty and now you have to do it all at once. Chores first, then TV."

Maddie and Justin slowly got up from the floor and went to dress for their job. They knew she was right, but somehow, something always came up each day that was more important than bringing in wood. They walked out into what seemed to be the North Pole, shivering as the wind blew past them and swirled snow at their feet.

"C'mon," said Justin. "Let's get this over with. We have a huge box to fill."

An hour later, two very rosy-cheeked children put the last log into the box. They were tired but the exercise had made them feel warm.

"There, that's done for a while," said Justin as he closed the wood box lid. "From now on, I'm carrying in my three logs every day. Filling the whole box is more than a chore!"

The children sat down after removing their coats, hats, and mittens. Now for a cozy hour with the cartoons! Maddie sat on her cushion near the TV and Justin on his, near the wood box. After about five minutes, Justin began to shiver.

"Hey!" he said. "It's cold down here!"

"You'll warm up," said Maddie. "You've been outside too long."

Another five minutes later Justin complained again.

"It's colder than ever!" he claimed.

Maddie scooted over to where he sat. She could feel the difference almost at once.

"It's beginning to get cold over by my spot too," she said. "Did you leave a door open?"

Justin checked. All the doors were closed.

"Well, where's that draft coming from?" asked Maddie.

The area around the wood box was definitely chilly. Too chilly for comfort. The call for lunch sounded from upstairs and the two trudged up for their midday meal. Later that afternoon, they came downstairs to a lovely warm room.

"I wonder what made it so cold this morning," Justin pondered. "Now

it's fine again."

"Maybe it was the wood," Maddie replied. "But how could wood make it chilly? Wood is supposed to make things warm."

"Yeah, when it's burning," explained Justin. "But what about when it's just sitting there?"

"Well," said Maddie, "it's just sitting now, and it's not cold in here anymore. What's the difference?"

PURPOSE

Wood comes from trees, right? It is full of potential chemical energy that can produce heat when it is put into a stove or fireplace and burned. Yet, in this story, the Earth's bounty seems to produce a cooler room. What can be the cause? Thermodynamics, the branch of physics that deals with the conversion of various forms of energy from one to another, affecting things such as temperature, is the answer to this riddle, since the flow of heat goes from warmer to cooler objects in a closed system. Once again, this story parallels personal experience, since my children and I also brought in a large quantity of wood from the cold outdoors into a warm room. This story can lead to a great discussion of our uses of the words *warm* and *cold* in the colloquial sense as compared to the words physicists use. It can raise questions about the wood created with carbon from the atmosphere and the Sun's energy and its potential to "produce" heat or, in this case, to absorb heat. It can also provoke inquiry about the nature of heat, temperature, and how heat moves from one substance to another.

RELATED CONCEPTS

- Thermodynamics
- Energy
- Temperature
- Cooling and heating
- Energy transfer
- Heat
- Specific heat

DON'T BE SURPRISED

Children have many misconceptions about heat and temperature. They believe that the two concepts are the same thing. If they believe that wool mittens contain heat, why not the same misconception about wood? (See the article my colleague, Bruce Watson, and I wrote about this in the *Phi Delta Kappan*, 1990. This article can be found online at *www.exploratorium.edu/ifi/resources/workshops/teachingforconcept.html*.) Although children's experience with wood is mainly with its burning, wood is still seen as a source of heat and not a substance that will cool off a room. Yet a mass of wood possessing a very low temperature is a *heat sink* and absorbs heat energy just like any other cool mass in a warmer environment. Again, our use of colloquial language will tend to cause the children to talk in terms of "cold" as an entity rather than as a *lack* of heat energy. Experience has shown that a lively discussion will ensue after the story is read to the class.

CONTENT BACKGROUND

Our two happy children sitting in the cozy room were unaware of the reason for the warmth in their surroundings. The wood stove with logs alight was *radiating*

heat out into the cooler room and as far as the children were concerned, the room was warm enough for their comfort. *Radiation* occurred when the burning logs heated the metal stove, which gave off infrared rays that traveled both directly to the children close enough to the stove to receive them and into the air molecules in the room. The space between these air molecules began to expand as the molecules got warmer and rose toward the ceiling. These warm molecules were replaced by cooled air molecules tumbling down, which were in turn warmed by the radiation from the stove. Thus a circular current of air moved throughout the room, from floor to ceiling and back again, creating fairly even temperatures for our children in what is called a *convection cell*.

Then things changed as the children brought in a large mass of wood that had lost a great deal of heat in the freezing outdoors. It had become as cold as the air outside the house. One part of the Law of Thermodynamics states simply that *heat* energy flows from hot to cold. Even though the wood stove was doing its best to keep the room temperature constant, much of the warm air in the room moved toward the cold wood in sufficient amounts to cause a drop in the temperature and even cause what Maddie described as a draft of cold air. Finally, the room warmed up as the temperature difference between the once ice cold wood and the room air reached equilibrium, and since there was no difference between the temperatures of the two, heat only flowed enough to maintain this equilibrium. This flow of heat goes on continuously in any space because everything in that space must attain or maintain an equal temperature. In fact, *heat* is defined as the flow of energy from an object that is warm to one that is cooler.

Before discussing this with your class, I suggest that you read up on the subject of energy transfer in a physical science book. One good option is *Science Matters* by Hazen and Trefil (1991), specifically chapter 2 on energy. Its simple and down-to-earth explanations will give you confidence in your understanding of the concepts used here.

Basically, there are three terms that often are used incorrectly in everyday speech: *thermal energy, temperature,* and *heat. Thermal energy* refers to the total amount of *kinetic energy* in a substance caused by the motion and collisions of the molecules that make up that substance. *Temperature* is a term that refers to the average kinetic energy in a substance and is measured by an instrument called a thermometer, which uses any of the three arbitrary scales: Fahrenheit, Celsius, and Kelvin (named after the people who devised them). Each has its own values and uses but that is another story. And, as we said before, *heat* is the flow of energy from a warmer object to a cooler object.

Let's talk a moment about temperature and thermal energy. Consider a pot of boiling water. When you want to take the temperature of the water, you can put the thermometer only in one place. It will register only the thermal energy of the molecules that strike it at that place. You can assume that because of convection currents, the kinetic energy throughout the pot is fairly consistent. Therefore you would get essentially the same reading wherever you put the thermometer. This is the water's *temperature*.

But this is not a measure of the total amount of *thermal energy* in the pot of water. It is only a snapshot of the average kinetic energy of the molecules striking the thermometer at that particular place in the pot. A single drop of that water or the entire pot of water would have the same temperature, but pouring the entire pot over your hand and putting a single drop on your hand would result in very dissimilar outcomes. In other words, the amounts of *thermal energy* transferred to your hand would be seriously different. Even though this analogy is not one we would wish to demonstrate, it does point to the difference between temperature and thermal energy.

Back in the warm room and the addition of the ice-cold wood mass, where we can see that the *thermal energy* that made the room cozy for the children did not stand a chance against the effects of thermodynamics. The *temperature* difference between the huge mass of wood and the air surrounding it dictated the transfer of *heat* from the room air to the wood until they were the same temperature. When that equilibrium is reached the only flow of heat that occurred was that which was needed to reach equilibrium, and that would be more or less constant but not evident to our limited senses. Eventually, everything in the room, with the exception of the stove and the people, who are producing heat, would maintain the same temperature.

So we can see that any cold mass (regardless of the material it is made of) introduced into a warmer environment would produce the cooling effect felt by Justin and Maddie. Some materials do take in thermal energy and give it off at different rates. For example, water requires more energy and time to raise its temperature than many other substances and by the same token gives off that amount of energy more slowly. Metals, however, react in a different manner. It takes less energy to raise the temperature of most metals. They absorb heat faster and release that heat faster than water—or wood, for that matter. Thus, if you were to touch a log and a metal object which had the same temperature, you would find that the metal object felt cooler since it would take heat from your hand faster than the log and thus feel cooler. You may have noticed also that if you are heating water on a stove in a metal pan the pan gets too hot to touch before the water does.

reLateD IDeas From NatIonaL scIence eDucatIon stanDarDs (Nrc 1996)

K–4: Properties of Objects and Materials

- Objects have many observable properties, including size, weight, shape, color, temperature and the ability to react with other substances. Those properties can be measured using tools, such as rulers, balances, and thermometers.
- Objects are made of one or more materials, such as paper, wood and metal.
- Objects can be described by the properties of the materials from which they are made, and those properties can be used to separate or sort a group of objects or materials.

K–4: *Light, Heat, Electricity, and Magnetism*

- Heat can be produced in many ways such as burning, rubbing and mixing one substance with another. Heat can move from one object to another.

5–8: *Transfer of Energy*

- Energy is a property of many substances and is associated with heat, light, electricity, mechanical motion, sound, nuclear energy, and nature of a chemical change. Energy is transferred in many ways.
- Heat moves in predictable ways, flowing from warmer objects to cooler ones until both reach the same temperature.

related ideas from benchmarks for science literacy (aaas 1993)

K–2: *Energy Transformation*

- The sun warms the land, air, and water

3–5: *Energy Transformation*

- When warmer things are put with cooler ones, the warm ones lose heat and the cool ones gain it until they are all the same temperature. A warmer object can warm a cooler one by contact or at a distance.
- Some materials conduct heat much better than others. Poor conductors can reduce heat loss.

6–8: *Energy Transformation*

- Heat can be transferred through materials by the collision of atoms or across space by radiation. If the material is fluid, currents will be set up in it that aid the transfer of heat.
- Energy appears in different forms. Heat energy is in the disorderly motion of molecules.

USING THE STORY WITH GRADES K–4

You may be wondering why this story is placed in the Earth and space section of the book when the concepts seem to be more applicable to the physical sciences. My opinion is that the Earth sciences are so directly related to all other sciences that any discussion of energy in any form is applicable to them. This story deals with energy within the atmosphere and its concepts would fit into looking at energy transformations and transfer, a big conceptual aspect of Earth sciences.

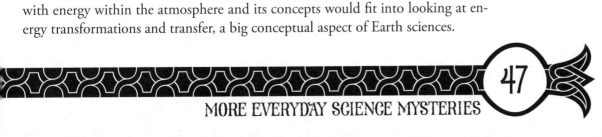

Particularly in the lower grades, where the concept of energy is not specifically addressed by the Standards or the Benchmarks, teaching about energy may present a problem. However, every young child has experienced the effects of higher or lower temperatures whether at home or away from home. Even at the early ages, children can be made aware of the existence of heat and know the difference between heat moving out of something and thereby lowering its temperature and "cold" moving in with the same result.

Helping K–2 students become familiar with a thermometer as a tool for measuring changes in material is a worthy goal. Keeping track of outdoor and indoor temperatures and keeping a weather log is also a wonderful way to connect science and communication skills. Michael Klentschy's book *Using Science Notebooks* (2008) has many suggestions for helping children keep records in science notebooks. Starting early on this sort of learning is important. He suggests that students have the opportunity to connect images and text and that observation can be recorded with either or both. Students can plot differences between temperatures by drawing thermometers even if they cannot read and understand the numbers. I have had success in using Centicubes to represent changes in temperature so that children can compare by placing one column next to another. This is, of course, beginning with a qualitative approach to comparison but for very young children it is often the best way.

With grades 3 and 4 there is usually no problem in using a thermometer and they can easily recreate the situation in the story. I suggest using a Styrofoam container large enough to place 2–3 logs that have been cooled by placing them in a freezer or, if possible, outdoors, if temperatures drop low enough in your area. An indoor/outdoor thermometer is a good way to measure temperatures, with the outdoor probe placed inside the container so that room temperature (which is where the temperature will start before the wood is added) and changes inside the container can be recorded. Predictions are in order here, and hypotheses can be placed on an "Our Best Thinking" chart or in students' science notebooks and kept for future reference.

If you use the method of recording the statements of the children about what they think they understand about the problem in the story on the "Our Best Thinking" chart, you can get statements that can be changed to questions. Questions may include

- Will the temperature change inside the box after the wood is added?
- How long will it take before the temperature stops changing?
- How will fewer or more logs affect the time and temperatures?
- What will happen to the temperature if the lid is removed?
- If the lid is removed, how long will it take before the temperature stops changing?
- What differences, if any, will we find if we use other materials?
- Does the amount of material make a difference in what happens?

Other questions may arise as the activities proceed so be prepared to modify the lesson as the questions dictate. This is, after all, what inquiry is all about.

USING THE STORY WITH GRADES 5–8

Students will probably suggest creating a simulation of the circumstances represented by Justin and Maddie for testing the accuracy of the story. You may suggest the use of the Styrofoam container if they do not. It will probably bring up the idea of insulation and insulating materials. The room the children were in was probably insulated from the outside since that is the building code in most states and is a way of saving on heat bills as well. This brings up an opportunity to talk about why we insulate our homes against heat entering or leaving our living quarters.

I would suggest using the statements of the children about what they predict will happen on the "Our Best Thinking" chart and saving the chart and changing the statements to questions. The questions will probably be similar to those listed above in the grades 3–4 section, but there may be more involving other materials and keeping the variables the same and controlling for differences. The discussions about experimental design will probably be heated since it will be difficult to find materials that are exactly the same mass to test against each other. They should test their hypotheses about the amount of temperature differences expected due to differences in mass.

Again, the inquiry involved here is the goal. Planning and carrying out investigations, which involves interpreting data and drawing conclusions, gives the students a chance to engage in real science. Remember, the story is a stimulus to bring out the students' misconceptions about heat and temperature and allow them to own their investigations. They should have no trouble finishing the story of Maddie and Justin after exploring the many facets of thermodynamics in the atmosphere of the container.

RELATED NSTA BOOKS AND JOURNAL ARTICLES

Damonte, K. 2005. Heating up and cooling down. *Science and Children* 42 (8): 47–48.

Driver, R, A. Squires, P. Rushworth, and V. Wood-Robinson. 1994. *Making sense of secondary science: Research into children's ideas.* London and New York: Routledge-Falmer.

Keeley, P. 2005. *Science curriculum topic study: Bridging the gap between standards and practice.* Thousand Oaks, CA: Corwin Press.

Keeley, P., F. Eberle, and L. Farrin.2005. *Uncovering student ideas in science: 25 formative assessment probes* (vol. 1). Arlington, VA: NSTA Press.

Keeley, P., F. Eberle, and J. Tugel. 2007. *Uncovering student ideas in science: 25 More formative assessment probes* (vol. 2). Arlington, VA: NSTA Press.

Keeley, P., F. Eberle, and C. Dorsey. 2008. *Uncovering student ideas in science: Another 25 formative assessment probes* (vol. 3). Arlington, VA: NSTA Press.

Klentschy, M. 2008. *Using science notebooks in elementary classrooms.* Arlington, VA: NSTA Press.

May, K., and M. Kurbin. 2003. To heat or not to heat. *Science Scope* 26 (5): 38.

references

American Association for the Advancement of Science (AAAS). 1993. *Benchmarks for science literacy:* New York: Oxford University Press.

Hazen, R., and J. Trefil. 1991. *Science matters: Achieving scientific literacy.* New York: Anchor Books.

Klentschy, M. 2008. *Using science notebooks in elementary classrooms.* Arlington, VA: NSTA Press.

National Research Council (NRC). 1996. *National science education standards.* Washington, DC: National Academy Press.

Watson, B., and R. Konicek. 1990. Teaching for conceptual change: Confronting children's Experience. *Phi Delta Kappan* 71 (9): 680–685.

CHAPTER 6
THE NEW GREENHOUSE

Eddie and Kerry's mom is a master gardener. She takes care of other peoples' gardens and raises plants from seeds and from what she calls cuttings to sell to her customers. This year she had a new building installed in the yard to help her expand her business. It was called a solar greenhouse. It was about 8 feet (2.5 meters) wide, 10 feet (3 meters) long, and about 10 feet (3 meters) high at the peak. It looked like a one story house with glass walls and glass roof, two windows up on the top that could be raised or lowered, and a door. It took a lot of time and a lot of people to put it together, but when it was finally up, it looked great. They put in a water hose and electrical outlets. Eddie and Kerry's father built a bunch of tables inside with screen tops so that Mom could water her seedlings, letting the water seep through onto the gravel on the floor.

In the early spring of the year, their mother planted a lot of seeds in little trays and put them in the greenhouse. She had never had a greenhouse before and had a lot to learn about how they work. She knew that the Sun was supposed to shine down through the windows and warm the inside even though it was chilly outside. Somehow, the heat that came in through the windows stayed inside the greenhouse and didn't go back out. The windows were sealed, the door was tight, and the Sun shone brightly on the greenhouse. During the day in an unusual spring warm spell, the temperature got so hot that the windows and doors had to be opened to keep the temperature from frying the little seedlings. She even put a fan in the door to bring in some cool air to keep the temperature at a reasonable level.

Evening and overnight was a different matter however. As the temperature outside dropped and the Sun set, the temperature in the greenhouse was no more than a few degrees warmer than the outside air and, in early spring in New England, that was near or at freezing. Those little seedlings were in trouble.

"I am at a loss about how to keep the climate in my new greenhouse at a level that is good for the plants!" said Mom.

"Why don't you look on the internet and find out about greenhouse care?" said Kerry.

That's exactly what she did. She found out that in her case, she was dealing with what was called a passive solar greenhouse, or a greenhouse that was strictly at the mercy of the Sun. Commercial greenhouses had heaters and sprinklers, automatic vent raisers, and all sorts of gadgets much too expensive for Mom's business. One article suggested putting large rocks in the greenhouse to soak up heat during the day and give it back off during the night.

"Where am I going to get huge rocks and if I get them, where am I going to put them so there is still room for the plants?"

"Maybe we can use something other than rocks to collect heat," said Eddie. "I read that different things take in heat and give it off, either quickly or slowly. Don't remember which ones though. I also heard that dark colors make a difference, but I am not so sure that makes sense."

"I guess we can try some things. I have buckets for water or some other liquid we could try. Maybe that would work," said Mom.

"But water is cold. Do we have to use warm water?" Eddie asked.

"Well, I don't really know. But I do know it's going to take a lot of experimenting and keeping records," said Mom. "Everybody up for that?"

PURPOSE

In our economy, with the recent oil crisis and the problem of fuel supply versus insatiable demands, the importance of alternative energy sources is abundantly clear to anyone who has been to a gas pump or seen an electric bill lately. Those of us concerned about the health of the planet are sensitive to the differences between energy sources and how we use them and the disparity in this among poor and wealthy countries. Young children may be aware only of what they hear from adults, while older children may even be involved in organizations that are trying to make these problems known to the public. All of us, young and old, are now conscious of the differences it is making in our lifestyles, be it in concerns about automobiles and miles-per-gallon ratios or the need to take fewer trips due to gas costs. Developing an alternative energy project can make all of us more aware of the finite aspects of the current sources of power.

The technological aspects of this story are also important in that students can use their new learning about energy supply and storage for the development of a more efficient product. This is putting science to use in solving a real-life problem.

RELATED CONCEPTS

- Technology
- Alternative energy
- Renewable energy
- Solar energy
- Energy transformation
- Thermodynamics
- Conservation of energy

DON'T BE SURPRISED

Your students may not realize that the Sun's rays differ in their angle during the year. Many will still believe that the reason for summer and winter is the distance of the Earth from the Sun. They do not realize that it is the angle of the Sun striking the Earth that causes seasons. Yet the angle and direction of the Sun during the day is important in heating solar greenhouses. Many students are also not aware of the differing heat absorption and radiation properties of materials. They may be unaware of the terrific amount of heat that can be captured by a transparent building and be amazed by the great disparity between outdoor and indoor temperatures on sunny days. Some may even be surprised that solar energy can be "stored" in substances like water or rocks.

CONTENT BACKGROUND

This story is based upon using solar energy and a variety of materials to modify and channel this energy to capture and hold heat. Almost everybody has experienced the differences in temperature due to sunshine passing through windows into an

enclosed space. Solar light is composed of many wavelengths, ranging from the sub-spectrum invisible level through the visible spectrum and beyond to invisible ultraviolet ranges and above. This means that light comprises low-energy to high-energy ranges. Scientists call this *electromagnetic (EM) radiation*. In some cases this radiation behaves like waves and in others, like particles. (Quantum mechanics has attempted to unify these two models of EM but we won't go there in this book! Let's just say that scientists have agreed that science can live with seemingly contradictory explanations and that sometimes it makes sense to explain EM radiation as wave behavior and sometimes as particle behavior.) In talking about solar energy, it is most useful to talk about EM radiation as waves. There are radio waves that have low energy, and gamma rays that have high energy. In between, from low to high, are microwaves, infrared waves, our visible spectrum, ultraviolet waves, and x-rays.

When light from the Sun enters an enclosed area by way of a transparent passageway, such as a window or transparent or translucent plastic, it enters in a short-wave and high-energy state. Objects in the space, such as dirt, wood, and water, absorb it. The light is then reradiated from these objects as lower-energy waves which do not readily pass out of the clear areas through which the light entered. In essence, it is "trapped" in the enclosed space, and that space heats up as a result. This is the famous "greenhouse effect." We are all familiar with it if we have ever stepped into a car that has been sitting in the sun with the windows rolled up. It can be cool outside, but the car is much warmer than the outside air.

According to some experts, if you have in the greenhouse approximately two gallons of water for every square foot of glazing, you can maintain a temperature about 30°F above the outside temperature. Massive material, like rocks or concrete helps greatly, but pound for pound, water in closed, black containers is the best retainer of heat and the best way to keep temperatures in greenhouses stable. The reason for this is that water is slow to store heat and just as slow to transfer it. Different substances have what is known as *specific heat*, meaning they absorb and transfer heat at different rates. Metals need less heat to raise their temperatures and transfer heat very quickly as well. Rock or concrete needs more heat to raise its temperature, and water needs the most. The concrete paving and buildings in cities helps to keep city streets warm on sunny days, and the oceans and lakes have a moderating effect on temperatures in seaside or lakeside communities.

Our experience in our passive solar greenhouse is that ventilation is necessary on very sunny days since the temperature can reach over 100°F by late afternoon, which will likely kill the plants inside. So vents in the roof allow the warm air to escape into the cooler atmosphere. A fan is often also necessary to keep the temperature from rising to dangerous levels.

All of this is pertinent to the greenhouse effect so prevalent in the news today, since many scientists and climatologists believe that our society is producing what are known as "greenhouse gases" in such great amounts that we are in essence preventing heat energy from escaping from the Earth as it used to, thus causing the temperature of the atmosphere to rise. The gases are forming a barrier in the atmosphere, just like the glazing in a greenhouse or the glass in a car, and preventing the release of heat back into space. Thus the average temperature of the Earth is rising. The term for this is, of course, "global warming."

With this story, students can explore the effects of solar energy, energy absorp-

NATIONAL SCIENCE TEACHERS ASSOCIATION

tion, and energy dissipation of various substances and also the effect of dark and light colors on the absorption of heat in closed systems.

related ideas from National science education standards (NrC 1996)

K–4: Objects in the Sky

- The Sun provides the light and heat necessary to maintain the temperature of the Earth.

K–4: Identify a Simple Problem

- In problem identification, children should develop the ability to explain a problem in their own words and identify a specific task and solution related to the problem.

K–4: Propose a Solution

- Students should make proposals to build something or get something to work better; they should be able to describe and communicate their ideas. Students should recognize that designing a solution might have constraints, such as cost, materials, time, space, or safety.

5–8: Transfer of Energy

- Energy is a property of many substances and is associated with heat, light, electricity, mechanical motion, sound, nuclei, and the nature of a chemical. Energy is transferred in many ways.
- Heat moves in predictable ways, flowing from warmer objects to cooler ones, until both reach the same temperature.
- The Sun is a major source of energy for changes on the Earth's surface. The Sun loses energy by emitting light. A tiny fraction of that light reaches the Earth, transferring energy from the Sun to the Earth. The Sun's energy arrives as light with a range of wavelengths, consisting of visible light, infrared, and ultraviolet radiation.

5–8: Understanding About Science and Technology

- Science and technology are reciprocal. Science helps drive technology as it addresses questions that demand more sophisticated instruments and provides principles for better instrumentation and technique. Technology also provides tools for investigations, inquiry, and analysis.

related ideas in benchmarks for science literacy (aaas 1993)

K–2: Design and Systems
- People can uses objects and ways of doing things to solve problems.

K–2: Energy Transformations

- The Sun warms the land, air, and water.

3–5: Design and Systems
- There is no perfect design. Designs that are best in one respect may be inferior in other way. Usually some features must be sacrificed to get others.

3–5: Energy Transformations
- Things that give off light often also give off heat.
- Some materials conduct heat much better than others. Poor conductors can reduce heat loss.

6–8: Design and Systems
- Design usually requires taking constraints into account. Some constraints, such as gravity or the properties of the materials to be used, are unavoidable.

6–8: Processes That Shape the Earth
- Human activities, such as reducing the amount of forest cover, increasing the amount and variety of chemicals released into the atmosphere, and intensive farming, have changed the Earth's land, oceans, and atmosphere. Some of these changes have decreased the capacity of the environment to support some life forms.

6–8: Energy Transformations
- Energy cannot be created or destroyed, but only changed from one form into another.
- Most of what goes on in the universe—from exploding stars and biological growth to the operation of machines and motion of people—involves some form of energy being transformed into another. Energy in the form of heat is almost always one of the products of an energy transformation.
- Energy appears in different forms. Heat energy is in the disorderly motion of molecules.

USING THE STORY WITH GRADES K-4

With thanks to Peggy Ashbrook (2007), I suggest that very young students begin their inquiry into this story outdoors on a sunny day with a discussion about what they can sense about the light coming from the Sun. Moving the students around from an absorbing surface like asphalt to more reflective surfaces may let them feel the differences in the heat. Making "Sun prints" on light-reactive paper using incandescent light and sunlight and then comparing them may give the children a sense of the power of the Sun's energy and its effect upon them and the world around them. Connecting this to the story can be done by asking students to relate how the energy from the Sun could cause the greenhouse to heat up. Since the students have developed an appreciation for the power of the Sun's energy, they may well be able to relate the two situations but not much more at this age.

When working with grades 3 and 4, I believe it would be advantageous to focus on the absorbing function of dark colors and the reflecting function of light colors before considering whether or not to address the greenhouse effect, which is the basis of the story. This can be done by placing thermometers on dark surfaces and on light surfaces and comparing temperatures. Ask them if they have noticed that the color of clothes they wear on really sunny days affects how hot they feel. This should lead to a discussion about the amount of solar energy the dark-colored clothing absorbs or light-colored clothing reflects. Then, investigations can begin with thermometers and various colors of T-shirts or baseball caps.

Once you have determined if the students are aware of the absorbing and reflective attributes of objects, they may be able to move into the greenhouse and realize that the soil and other objects in the greenhouse have been able to absorb the Sun's energy and hold it inside the structure. A field trip to a local botanical garden may be in order. It would be surprising if the students at this level were able to apply their knowledge of solar energy to the technological aspects of the greenhouse. But it is possible, especially with older students or ones who may have had experience with enclosed solar rooms, either in their own homes or in family businesses. Read the following segment on using the story with grades 5–8 to see if your students are up to the complexities of investigating some of the questions that arise.

I also recommend you read the Damonte article in *Science and Children* "Heating Up and Cooling Down" (2005), which will be very appropriate, even for younger children.

USING THE STORY WITH GRADES 5-8

This story lends itself to a great deal of scientific and technological investigation for children at this grade level. Gregory Childs, in his *Science and Children* article, "A Solar Energy Cycle" (2007), suggests checking out whether or not your students are aware of the absorbing and reflecting power of light and dark surfaces first before moving into the greenhouse investigations. He uses two paper houses, one black and one white, placed in the sun to compare temperatures. They need not be works of art—simple cardboard boxes should do nicely. The results of this investigation will be an imbedded formative assessment for you to see what your students understand about this concept.If you are comfortable with what you find, you can move on to the ideas involved in the story about the greenhouse effect.

The story itself gives clues to the differences in heat absorption of various materials, and controlled experiments can be conducted using greenhouses constructed by your students using cardboard boxes and plastic wrap or more authentic balsa wood greenhouse constructions, depending upon the sophistication of your students. Many students have built greenhouses in the 12″ ×14″ size, many complete with vents in the upper windows using duct tape for hinges, etc. My own attempt at a very large plastic solar greenhouse built by my class on the roof of my school was thwarted when, during a wind storm, it became a wonderfully aerodynamic box kite, sans line, and ended up several blocks away from the school in someone's backyard! Luckily no property or bodily damage resulted, and the recipients of the structure were friendly and understanding. But I learned a great deal about size, wind, and the importance of anchoring. For this reason, I do suggest keeping your projects small and portable so that they can be taken indoors and out of the unpredictable elements.

There are a few variables to be tested here, for example, the effect of dark and light surfaces in the greenhouse; the differences in storage mass material; the effects of insulation; and most basic of all, witnessing the simple solar greenhouse effect in order to establish a base line set of data. You may be able to round up some of the black plastic film canisters that, with lids, make wonderful water, rock, sand, or gravel containers. It is good to remember that we are talking about mass here, and controlling the mass is important when comparing the various materials.

On sunny days, exposures for 20 minutes are usually sufficient, with a reading every 2 minutes or so. This gives a great deal of data to be graphed and analyzed. If you are fortunate to have a temperature probe and graphing program, this can be used as well. If your students can procure digital probe thermometers, it makes reading easier, but if not, a thermometer placed near a window of the greenhouse can be read without opening the structure. You can also test color absorption and temperature effects by using different color floors in the greenhouses. This has proved to be very effective in producing significant data. With guidance, your students will be able to make a list of significant variables that must be controlled and the results should be rewarding both from a conceptual understanding and a technological standpoint.

Of course, no project on solar energy is complete without a discussion of alternative energy. With oil and gas prices fluctuating wildly and these fossil fuels being depleted, solar energy is poised to become a major alternative. Some schools have hosted simulated symposia and/or town hall meetings to discuss this issue. The classroom setting can be very conducive for impressionable students to become aware of the seriousness of this issue as it relates to their future.

related nsta books and journal articles

Ashbrook, P. 2007. The early years: The sun's energy. *Science and Children* 44 (7): 18–19.

Childs, G. 2007. A solar energy cycle. *Science and Children* 44 (7): 26–29.

Damonte, K. 2005. Heating up and cooling down. *Science and Children* 42 (8): 47–48.

Driver, R., A. Squires, P. Rushworth, and V. Wood-Robinson. 1994. *Making sense of secondary science: Research into children's ideas.* London and New York: Rout-ledge-Falmer.

Keeley, P. 2005. *Science curriculum topic study: Bridging the gap between standards and practice.* Thousand Oaks, CA: Corwin Press.

Keeley, P., F. Eberle, and L. Farrin. 2005. *Uncovering student ideas in science: 25 formative assessment probes* (vol. 1). Arlington, VA: NSTA Press.

Keeley, P., F. Eberle, and J. Tugel. 2007. *Uncovering student ideas in science: 25 more formative assessment probes* (vol. 2). Arlington, VA: NSTA Press.

Keeley, P., F. Eberle, and C. Dorsey. 2008. *Uncovering student ideas in science: Another 25 formative assessment probes* (vol. 3). Arlington, VA: NSTA Press.

May, K., and M. Kurbin. 2003. To heat or not to heat. *Science Scope* 26 (5): 38.

references

American Association for the Advancement of Science (AAAS). 1993. *Benchmarks for science literacy.* New York: Oxford University Press.

Ashbrook, P. 2007. The early years: The sun's energy. *Science and Children* 44 (7): 18–19.

Childs, G. 2007. A solar energy cycle. *Science and Children* 44 (7): 26–29.

Damonte, K. 2005. Heating up and cooling down. *Science and Children* 42 (8): 47–48.

Hazen, R., and J. Trefil. 1991. *Science matters: Achieving scientific literacy.* New York: Anchor Books.

Klentschy, M. 2008. *Using science notebooks in elementary classrooms.* Arlington, VA: NSTA Press.

National Research Council (NRC). 1996. *National science education standards.* Washington, DC: National Academy Press.

CHAPTER 7
ROTTEN APPLES

Ted and Steve were on their way home from school one October day and decided to take a short cut, as they often did, through the old apple orchard near their homes. "You know, Steve, all of our homes are built on land that used to be part of a huge apple orchard. The owner sold off a lot of his old orchard to developers who built these homes. My mom remembers when all of this land, including our school, was apple orchard."

"Yeah, I remember hearing about that too," said Steve.

"I wonder what happened to all of the old apples that were all over this field and around our homes since the trees were cut down," said Ted.

"And look around this orchard now," said Steve, "There are all kinds of apples lying on the ground that didn't get picked or just fell after the harvest."

"You'd think that the apples from all of the years that trees have been dropping them would be around here and we'd be up to our knees in old apples," laughed

Ted. "But actually, I wonder where they all go to. By next spring the ground will be clear and all of these apples will be gone. Does someone come out here and clean them all up?"

"I don't think so, unless they do that to make apple cider," replied Steve. "Yeah, that's probably what happens, or else in the spring we'd be walking over a lot of apples."

It was a nice warm fall day and the boys took their time walking home and stopped to look at some of the apples left on the ground.

"Man, look at these apples," said Ted, "They look half-rotten already. They wouldn't make cider out of these would they? They're all goopy and look like they have worms or something in them. They're not good for anything any more."

"I'll bet animals eat a lot of them but not all of them, 'cause there are still a lot of them left," offered Steve. "So why aren't they here in the spring? And where are the apples around our school and our houses?"

"I heard that they turn into soil," said Ted.

"Just like that? Magic?" asked Steve. "How can that be, soil is soil and it's always here, with or without apples. Soil is dirt, right, so apples can't turn into dirt! It has to be more complicated than that."

"I know," said Ted, "Let's take some of them home and put them in my yard and see what happens. We'll put them outside where the dogs can't get at them and we'll keep an eye on them. Nothin' like good ole observation, like Ms. Green keeps telling us in science class."

"Okay, as long as it's in your yard," said Steve, "I don't think my folks will like rotten apples in my yard and anyway our dog would eat them for sure. She eats anything, and I mean *anything*!"

And so the boys took a bunch of apples from the ground and took them home for some "good ole observation," just like Ms. Green kept telling them. And next spring…?

PURPOSE

In 1991 and 1992, John Leach, Bonnie Shapiro, and I did a study in which we interviewed approximately 400 students from the United Kingdom, Canada, and the United States about their beliefs surrounding the decay of an apple over a year's time. We found that children in all three countries had little understanding of the process of decay and almost totally ignored the role of microorganisms in the process. They believed that apples or any other previously living thing left on the ground miraculously turned to soil, were eaten by animals, or merely disappeared. Therefore the main purpose of this story is to help students explore how decay breaks down organic material so that it can be recycled within its ecosystem.

DON'T BE SURPRISED

While interviewing a high school student who had just finished Biology I for the 1992 study (Leach, Konicek, and Shapiro), I asked her if she thought that the materials from a rotting apple could possibly be used by another plant for its nutrients. She looked me as though I was completely addled. Then she said, "Wait a minute! Water recycles…! Ah, but not real stuff like apples."

This is typical of children (and many adults) who believe that matter just disappears into the soil or becomes soil. The idea of decaying material becoming part of the nonliving environment is completely foreign. Because many young (and older!) people think that bacteria are simply disease-producing organisms, they suppose that ridding the world of all bacteria would be beneficial. They lack knowledge of decomposers in the soil that help break down dead organic material. They probably do not understand the cycle of conservation of matter and recycling of materials into new growth. With the recent media buzz about composting and environmental concerns, children may have become more aware of recycling, but don't be surprised if they are unaware of how the process takes place.

Also, some of your students might believe that the Earth is getting heavier each year because of the leaves that fall and fruits of the field that are left behind. Of course, the gist of the story you have just read, put simply, is that the planet reclaims its own material, and uses it over and over again.

RELATED CONCEPTS

- Decay
- Decomposition
- Decomposers microbes
- Fungi
- Oxidation

CONTENT BACKGROUND

July 5, 2008: Today, in the cool of the morning, we picked up a truckload of bark mulch for use in our garden and watched as vapor was emitted from the huge

pile each time the front loader scooped up a load to deposit in the bed of our truck. Inside that pile of bark mulch, tiny microbes were busy breaking down the organic material into its component parts: molecules of carbon and nitrogen and phosphorus-based compounds; and in the process, oxygen was being utilized and the oxidation process was releasing the energy stored in the plant material as heat. As the moist vapor condensed in the cooler air, little clouds appeared. I buried my hand in the bark mulch and could feel the heat, which I judged might have been as warm as 60°C. Indeed, in the field behind our home, the nearby college agriculture department often dumps their composting pile, and in the wintertime when the air is really cool, the clouds of vapor rising from it look almost like a steam vent.

What a wonderful process this is! Decomposition of organic matter by the microbes and fungi in the air and soil takes care of all of our unwanted organic material and returns the building blocks of their bodies to the earth to be used again and again. The same thing happens in the compost pile in our garden. We feed our vegetable garbage to the microbes and fungi, and in return, they give us beautiful, black, fertile soil to use in our garden. What a relationship we have with our tiny friends! We each have our jobs to do, by which we both benefit and, in turn, create a healthier planet.

In our compost pile each summer, we say hello to hundreds of worms, pill bugs, centipedes, and millipedes as they wiggle away from our intruding pitchfork while we turn their dinner table over to provide more oxygen and water. These creatures too are important, although they are not technically decomposers. Scientists call them *detritivores* or sometimes just scavengers. They do not completely digest all of the potential nutrients nor do they release all of the energy. The worms and other animals eat the garbage we put in for them and perform the first part of the decomposition process by transforming the large pieces of grass clippings, lettuce, avocado pits, and other plant material into more "bite-sized" pieces for the real decomposers, the microbes and fungi. By passing the unused portion of their meals through their bodies in a partially digested form, they speed up the process of composting. They also aerate the pile so oxygen and water can circulate and do their part in the process.

Ted and Steve will witness this phenomenon as they watch the apples over the fall and winter. The cold temperatures may slow the process down a bit, but sooner or later, the apples will become smaller and smaller as the decomposers use the food in the apples to satisfy their needs and release the essential nutrients back into the soil.

According to the U.S. Environmental Protection Agency nearly 25% of the solid waste that goes into landfills is made up of our organic garbage and lawn clippings. Almost all of this could be composted into usable material. The EPA website provides lists of desirable and undesirable materials for use in composting (*www.epa.gov/epawaste/conserve/rrr/composting/index.htm*).

related ideas in national science educations standards (NRC 1996)

K–4: The Characteristics of Organisms
- Organisms have basic needs.

K–4: Organisms and Their Environments
- All organisms cause changes in the environments where they live. Some of these changes are detrimental to the organism or to other organisms, whereas others are beneficial.

5–8: Populations and Ecosystems
- Decomposers, primarily bacteria and fungi, are consumers that use waste materials and dead organisms for food.

related ideas in benchmarks for science literacy (aaas 1993)

K–2: Flow of Matter and Energy
- Many materials can be recycled and used again, sometimes in different forms.

K–2: Constancy and Change
- Things change in some ways and stay the same in some ways.

3–5: Interdependence of Life
- Insects and various other organisms depend on dead plant and animal material for food.
- Most microorganisms do not cause disease and many are beneficial.

3–5: Flow of Matter and Energy
- Some source of energy is needed for all organisms to stay alive and grow.
- Over the whole Earth, organisms are growing, dying, and decaying and new organisms are being produced by the old ones.

6–8: Interdependence of Life

- Two types of organisms may interact with one another in several ways: They may be in a producer-consumer, predator-prey, or parasite-host relationship. Or, one organism may scavenge or decompose another.

6–8: Flow of Matter and Energy

- Food provides molecules that serve as fuel and building material for all organisms.

USING THE STORY WITH GRADES K–4

It might be a good idea before you read the story to your class to give them the probe "Rotting Apple" in Keeley et al.'s *Uncovering Student Ideas in Science: Another 25 Formative Assessment Probes,* (2008). In this probe, four friends argue about why an apple, over time, disappears from view. Students have to choose from one of the six possible arguments. For young children, you may want to narrow the number of arguments so as not to confuse them. Regardless, you will probably find out, as we did in our 1992 study, that children do not consider small organisms capable of using the energy in the apple to break it down. Further, your students will not be aware of the particulate nature of matter and so do not think of the apple as being composed of particles so small that they can be absorbed by the soil and used again. But they may think about the possibility of small organisms using the apple for nourishment, and this is a first step in recognizing the importance of life forms smaller than the worms and insects that they may have seen eating rotting apples. Creating the "Best Thinking" chart to ask for their expectations about the future of an apple placed in a transparent container with some soil will also give you some idea as to what your children are expecting.

Cover the container so that mold spores do not escape into the classroom atmosphere. The children will see the apple begin to rot and observe the mold growing on the apple as it slowly shrinks into mush. It is a good idea to break the skin slightly so that the soil organisms can gain entrance. A bruised apple will also lend its own enzymes to the process. Decay would happen regardless, but bruising or scoring the fruit will speed up the process for impatient youngsters. Drawings and notes should be recorded in student science notebooks and a summary put on the class chart so that all can see a daily or weekly record.

In the end, the apple will seem to dissolve into the soil with possibly a piece of skin and a stem left behind. Since the container was covered, the students think that the apple has become part of the soil. They may also have never noticed the mold that grows and although they cannot see the bacteria, you could suggest to them that there are other organisms too small to see that are working along with the mold.

USING THE STORY WITH GRADES 5–8

Use the probe mentioned in the section above as it is written. The students will probably suggest that they be allowed to do the same thing that Steve and Ted do in the story and follow it up with some "good ole observation." You might consider asking your students to suggest variables in their setups and have several stations for observing apples. These might include the following:

- Types of apples
- Sizes of apples
- Apples with bruises
- Cut apples
- Apples in pieces
- Apples without soil in container
- Apples with soil in the container
- Soils from various locations
- Different fruits such as bananas, grapes, kiwi, citrus.
- Comparison with leaves, grass clippings, and other common vegetation.
- Different temperatures
- Dry versus moist environments

Predictions can be listed at each of the various stations as to what the students think will happen in each case. Students will have to decide if the apples they use are to be gleaned from the ground or bought from a local orchard or grocery store. They should be concerned with keeping the variables at a minimum whenever possible. The website *www.break.com/index/fruit-decomposition-time-lapse.html* has a 30-second time-lapse video of two months of decomposition of fruit. Perhaps with today's ubiquitous technology, you and your students can set up your own time-lapse video of your decomposing apples. In case you might want to use the ever-popular bottle biology methods of investigating decomposition, you can visit their website at *www.bottlebiology.org.*

Your students may wonder how people help vegetables and fruits edible before the advent of refrigeration. The root cellar in most 17th- and 18th-century homes was used to store potatoes, carrots, and apples for use during the whole year. How did that work? Humidity and temperature were certainly factors but were there others? How did people keep the vegetables from freezing during the cold winters or rotting during warmer winters in warmer parts of the country? The reverse of learning about decomposition is learning how to preserve.

RELATED NSTA PRESS BOOKS AND JOURNAL ARTICLES

Driver, R., A. Squires, P. Rushworth, and V. Wood-Robinson. 1994. *Making sense of secondary science: Research into children's ideas.* London and New York: Routledge-Falmer.

Keeley, P. 2005. *Science curriculum topic study: Bridging the gap between standards and practice.* Thousand Oaks, CA: Corwin Press.

Keeley, P., F. Eberle, and L. Farrin. 2005. *Uncovering student ideas in science: 25 formative assessment probes.* Arlington, VA: NSTA Press.

Keeley, P., F. Eberle, and J. Tugel. 2007. *Uncovering student ideas in science: 25 more formative assessment probes* (vol. 2). Arlington, VA: NSTA Press.

Keeley, P., F. Eberle, and C. Dorsey. 2008. *Uncovering student ideas in science: Another 25 formative assessment probes* (vol. 3). Arlington, VA: NSTA Press.

references

American Association for the Advancement of Science (AAAS). 1993. *Benchmarks for science literacy:* New York: Oxford University Press.

Bottle Biology. *www.bottlebiology.org.*

Hazen, R., J. Trefil, 1991. *Science matters: Achieving scientific literacy.* New York: Anchor Books.

Keeley, P., F. Eberle, and C. Dorsey. 2008. *Uncovering student ideas in science: Another 25 formative assessment probes* (vol. 3). Arlington, VA: NSTA Press.

Klentschy, M. 2008. *Using science notebooks in elementary classrooms.* Arlington, VA: NSTA Press.

Leach, J. T., R. D. Konicek, and B. L. Shapiro. 1992. *The ideas used by British and North American school children to interpret the phenomenon of decay: A cross-cultural study.* Paper presented to the Annual Meeting of the American Educational Research Association. San Francisco.

National Research Council (NRC). 1996. *National science education standards.* Washington, DC: National Academy Press.

CHAPTER 8
NOW JUST WAIT A MINUTE!

The science challenge had been set and the whole school was excited about competing. This year, instead of the usual science fair, the science committee asked each student competing to meet a special challenge in any way they could think of.

They could work in teams or as individuals but they had to follow strict rules, included how much they could spend and what materials they could use.

The challenge was to build a timing device that would measure exactly one minute. It could not have any clock mechanisms in it and had to rely strictly on a force to make it work—any force the participants chose. It could not cost more than $5.00 in materials. Judging would be based on how close each team's timer came to measuring exactly one minute!

Enrique and Katerina wanted to work together but they had very different ideas on how to make the timer and what forces to use. Enrique wanted to use gravity and Katerina wanted to use buoyancy, the force that makes certain things float or sink.

"I want to make a ramp for a marble that takes one minute to roll down a maze," said Enrique. "It will use the force of gravity."

"That's too easy," said Katerina. "We ought to use a water clock. That's a bowl that has a hole in the bottom of it and floats in another bowl for only one minute and then sinks."

"I know what a water clock is but that's pretty easy too," said Enrique, "So maybe we can figure out something that is really cool and something nobody else would think of."

"We ought to be able to come up with something new and different. Anyway, if we can't think of anything, we can always fall back on the water clock."

"Or the ramp," added Enrique.

"Okay, or the ramp," replied Katerina without much excitement in her voice. "One way or the other we ought to be able to make something that will do its thing in just one minute. How many forces are there that we can use to make our timer work? And how do we get it to do whatever it does in exactly one minute?"

National Science Teachers Association

PURPOSE

This story obviously is aimed at the technology standards. Two simple timing devices are mentioned with the suggestion that more are possible. These can be improved to meet the challenge or other devices could be invented. Students are being challenged to either improve on an idea that already exists or to be creative and think of another type of timing device. To judge this challenge, it is a good idea to have a stopwatch that is accurate to at least tenths of a second. I have witnessed contests that were decided by just this interval! Students often become very creative—and competitive—in meeting this challenge.

related concepts

- Time
- Accuracy
- Forces
- Design

DON'T BE SURPRISED

Time, especially for young children, is merely something that passes between important events. Long-time intervals may be between holidays, while short-time intervals may be between recess and lunchtime. How many parents have heard the question "How much longer?" when traveling or waiting to go somewhere? Clocks may have a meaning for some young children but even more effective are the many devices available based on the hourglass design. Somehow, seeing sand pass through a hole from top to bottom gives them a sense of beginning and end. Many of them will have used these devices in games they have at home or school.

Time often seems something set by adults. Many children cannot fathom why they can't see their favorite TV show when they want to see it rather than at the scheduled time they feel to be arbitrary. Also, don't expect your young students to connect the astronomical time clock to the daily running of their lives. Setting the clocks forward or backward for daylight saving time rarely changes the meaning of time as far as they are concerned. This is even so for some adults who still believe that the "extra hour of daylight will not be good for the crops," (a quote from a friend of my grandfather, years ago).

CONTENT BackGround

The measurement of time may have begun at the dawn of our species' arrival on this planet. Certainly, early humans were aware of the distinction between night and day and probably worshipped gods whom they believed controlled these happenings. The patterns of the celestial bodies that were so important to their lives did not go unnoticed or unrecorded. We have evidence of the significance of such events as solstices and equinoxes in structures that still survive. Stonehenge in southern England was a celestial clock and perhaps a place of worship. The stones in Stonehenge are arranged to coincide with various celestial happenings, including the equinoxes and solstices. Early Anasazi engineers in Chaco Canyon in northern New Mexico also developed structures for marking celestial happenings over a thousand years ago. And of course the Aztecs and Mayans of South and Central America had celestial calendars as early as the 11th century, CE (common era). These calendars were remarkably accurate and predicted important holidays, festivals, and of course the significant agricultural events of the year.

It is easy to infer from existing structures that from the earliest civilizations on, time was connected to the periodic motion of celestial bodies such as the Sun, the Moon, and the various constellations that were visible at different times of the year. Early Egyptians used obelisks as sundials and even developed elaborate water clocks called *clepsydras* (water thieves), which marked the day of important periods. Almost 2,000 years ago, in the Chinese empire, water clocks of great complexity were built and used to measure the passing of time, both day and night.

When clocks were built they had to have certain properties, the most important of which was the *consistent, periodic, constant,* and *repetitive* action that could mark off equal increments of time. Due to issues such as pressure and changes in dynamics, this was no easy matter and many designs were attempted over the centuries. For example, as water from a container dripped into another container, the pressure of the water could change causing the water to drip more slowly as time went on. Our ancestors were creative and clever and invented many ways of measuring time as the need for more accuracy evolved.

In more recent times, during the 1760s, an Englishman named John Harrison developed a clock and watch that kept such excellent time that it could be used aboard ships to ascertain longitude for oceangoing travel. His watch was accurate to five seconds over a period of months. And as a result, the pocket watch was born and became the standard of timekeeping for hundreds of year, even among the poorer citizens of the world.

The history of how various civilizations have used and manipulated time is a fascinating story and one worth learning. Travel and commerce dictated the need for accurate timetables, time zones, and all of the time-related manifestations of rules and regulations. The science of astronomy and the pseudoscience of astrology, politics, and religion have all contributed to time measurement and manipulation over the centuries.

Now in the 21st century, the American government keeps the standard of time with a cesium atomic clock that is so accurate that it will not lose or gain a second in 60 million years! (What's that old joke about "close enough for government work"?) The clock uses the regular movement of cesium atoms to calculate the time intervals. It's much more complicated than that, but this is a short book! You can find that exact time on the internet at *http://nist.time.gov*, in case you want to set your watch. As you can see, we have come a long way from obelisks and clepsydras and circles of stones. But if you expect that time scientists and engineers are satisfied with an accuracy of one second in 60 million years, you have another think coming. They are constantly trying to improve their invention, just as people have been trying to improve timekeeping inventions since the dawn of civilization. We are therefore asking your students to participate in the fun of improvement and to join those scientists of the past who tried to develop timekeepers that kept accurate time.

Two timekeepers are mentioned in the story, the maze and the water clock. The maze is merely a platform that allows gravity to pull a marble down through a maze of baffles which slow it down so that it reaches the bottom at a prescribed time rather than just plummeting down in fractions of a second. The baffles are modified by trial and error so that the marble takes the desired amount of time to reach the bottom.

Water clocks can be of several designs but essentially work on the basis of dripping water or water entering through a hole at a constant rate. One clock is a bowl with a small hole in the bottom which is set in another bowl of water. The hole allows water to enter and eventually sinks the bowl. The rate of sinking is controlled by the size of the hole. Another modification is a bowl that drips water into another bowl until it sinks the second bowl. Again the rate is determined by the size of the hole.

There are timekeepers that are made from candles that burn at a reliably constant rate. I have seen candles placed on a seesaw-like balance so that the balance moves as the candle burns and changes the balance between the two ends.

I return to the common properties of the timekeeper mentioned before, a consistent, periodic, constant, and repetitive action that will measure the intervals of time required. These properties can be interpreted liberally, such as in the case of the maze. But the main point is that the timekeeper must work consistently and not change over time. Your students should be able to come up with a myriad of designs to compete with others for the most accurate timekeeper.

reLaTeD IDeas From NaTIonaL ScIence eDucaTIon STanDarDs (NrC 1996)

K–4: Abilities of Technological Design

- Identify a simple problem.
- In problem identification, children should develop the ability to explain a problem in their own words and identify a specific task and solution related to the problem.
- Propose a solution.
- Students should make proposals to build something or get something to work better: they should be able to describe and communicate their ideas. Students should recognize that designing a solution might have constraints, such as cost, materials, time, space, or safety.

5–8: Abilities of Technological Design

- Design a solution or product.
- Students should make and compare different proposals in the light of the criteria they have selected. They must consider constraints—such as time, trade-offs and materials needed—and communicate ideas with drawings and simple models.
- Implement a proposed design.
- Students should organize materials and other resources, plan their work, make good use of group collaboration where appropriate, choose suitable tools and techniques, and work with appropriate measurement methods to ensure adequate accuracy.
- Evaluate completed technological designs or products
- Students should use criteria relevant to the original purpose or need, consider a variety of factors that might affect acceptability and suitability for intended users and beneficiaries and develop measures of quality with respect to such criteria and factors; they should also suggest improvement and, for their own products, try proposal modification.

RELATED IDEAS FROM BENCHMARKS FOR SCIENCE LITERACY (AAAS 1993)

K–2: The Nature of Technology

- Tools are used to do things better or more easily and to do some things that could not otherwise be done at all. In technology, tools are used to observe, measure, and make things.
- When trying to build something or to get something to work better, it usually helps to follow directions if there are any or to ask someone who has done it before for suggestions.
- People alone or in groups are always inventing new ways to solve problems and get work done. The tools and ways of doing things that people have invented affect all aspects of life.

K–2: Designs and Systems

- People can use objects and ways of doing things to solve problems.

3–5: Designs and Systems

- Even a good design may fail. Sometimes steps can be taken ahead of time to reduce the likelihood of failure, but it cannot be entirely eliminated.

3–5: The Nature of Technology

- Throughout all of history, people everywhere have invented and used tools. Most tools of today are different from those of the past but many are modifications of very ancient tools.
- Any invention is likely to lead to other inventions. Once an invention exists, people are likely to think up ways of using it that were never imagined at first.

6–8: Designs and Systems

- Design usually requires taking constraints into account. Some constraints such as gravity or the properties of the materials to be used are unavoidable.
- Technology cannot always provide successful solutions for problems or fulfill every human need.

USING THE STORY WITH GRADES K–4

You may have to remind the students of ways they keep time when playing games. It would be very surprising if some, if not all, of the students did not have at least a little experience with sand timers. Depending upon their motor skills level, you might ask them if they can make something similar themselves using pill bottles or other containers and sand. Ask them how they might make the timer keep longer time intervals or shorter time intervals. They usually suggest bigger containers, or more or less sand, or sometimes a change in the size of the hole through which the sand flows.

Younger children probably do not have the small motor coordination necessary to build a maze and marble ramp. However, the toy ramp and marble games that some children possess might be modified to take different times to operate. One is available from Sense Toys with a ramp that can be changed for slower or faster action. Lego blocks can also be used to create ramps which can be modified by the children. Another idea would be to use car ramps, which usually come with lots of extensions and curves and can be adjusted easily. Children may think that heavier balls will take longer to roll down the ramp and this can give you an opportunity to test this hypothesis.

My experiences with fourth graders and older have been very successful. They often want to make the maze out of wood but finally realize that box cardboard will work just as well and keep the cost down. They realize that the tilt of the maze board has a great effect on the timing device. Some have suggested placing a metal cup at the end so the ball will drop, signaling the end of the run. They also enjoy designing a water clock and soon realize that modifying the hole in the bottom of the bowl with tape has an effect on the length of time the bowl stays afloat. Some with more creative and/or technological bents will try to make tipping devices and/or candles with markings. The latter may prove difficult to read and are usually abandoned.

One of the most ingenious devices I have seen involved a candle on one end of a balance that tipped as the candle burned and lost weight. The balance was placed next to a scale of times marked on a piece of paper. It worked for several trials, but then it toppled over—a little engineering drawback! You may well be amazed at the different ideas that emerge from your students.

Be sure to take this opportunity for the students to use their science notebooks to record the challenge as they see it, their plan, and of course their results and conclusions. A list of problems incurred during the testing should also be included along with a narrative of how they went about trying to solve each setback. Their problem-solving techniques often provide the best discussions about techniques used in engineering analysis and action. Inviting an engineer into the classroom to talk with the children can also be a rewarding experience, both for the class and the engineer. Some engineering companies belong to nationwide organizations that help classroom teachers promote engineering curricula. Lockheed-Martin, Cisco, and Intel are just some of the sponsors of such enterprises. Look them up on the Internet to find out how they can help you in advancing your technological curriculum.

In the March 2007 issue of *Science and Children* there are several trade books recommended by Christine Anne Royce and ideas for using these books for K–3 children. The books are: *Let's Try it Out, Towers and Bridges* and *Bridges: Amazing Structures to Design, Build and Test*. I heartily recommend your looking into this article and the following suggestions for use with K-3 children. Although it does not directly apply to the story, it may offer a nice introduction into children building new structures to apply to problems that need to be solved.

USING THE STORY WITH GRADES 5-8

First, I would suggest that you read the K–4 section including the part about engineers in the classroom. Having worked with these organizations, I find them to be sincere and very skillful in working with students and teachers.

Middle school students usually jump at a chance to meet a challenge such as the one described in the story. You and the students can make up a rubric for assessing the success of each project. This rubric would allow some leeway for ideas that did not win the challenge but were within an agreed-upon range of accuracy and used good problem solving techniques. There could also be a criterion for creativity and for the most unusual creation. Many of the devices mentioned above can be created as well as the oft-used "domino-effect timer" made by stacked dominos knocking each other over for a given amount of time.

Another possible diversion can be a Rube Goldberg (1883–1970) contest, named for the cartoonist of the mid-20th century who was famous for creating depictions of complex devices for completing simple tasks. An excellent example of a Rube Goldberg type of device is in the famous Honda commercial which can still be found on YouTube. Encourage your students to use their search engines to see videos of Rube Goldberg devices and stimulate their creativity to use a combination of forces to accomplish a task. A book of his cartoons called: *Rube Goldberg: Inventions!* by M. F. Wolfe (2000) is a great resource for the students.

The website *rubegoldberg.com* is also the site for national contests and more information about his work. You can even find him in the dictionary as a noun. Look it up and see! As for the educational value of these forays into what may seem like fantasy, the planning and fine tuning of these devices are examples of what engineers have to do to accomplish their jobs while adding the whimsical and fun aspects to the challenge. It also allows students to use their knowledge of forces in unusual ways.

related NSTa Books and Journal articles

Driver, R., A. Squires, P. Rushworth, and V. Wood-Robinson. 1994. *Making sense of secondary science: Research into children's ideas.* London and New York: Routledge-Falmer.

Keeley, P. 2005. *Science curriculum topic study: Bridging the gap between standards and practice.* Thousand Oaks, CA: Corwin Press.

Keeley, P., F. Eberle, and L. Farrin. 2005. *Uncovering student ideas in science: 25 formative assessment probes* (vol. 1). Arlington, VA: NSTA Press.

Keeley, P., F. Eberle, and J. Tugel. 2007. *Uncovering student ideas in science: 25 more formative assessment probes* (vol. 2). Arlington, VA: NSTA Press.

Keeley, P., F. Eberle, and C. Dorsey. 2008. *Uncovering student ideas in science: Another 25 formative assessment probes* (vol. 3). Arlington, VA: NSTA Press.

Klentschy, M. 2008. *Using science notebooks in elementary classrooms.* Arlington, VA: NSTA Press.

Royce, C. A. 2007. If you build It. *Science and Children* 44 (7): 14–15.

references

American Association for the Advancement of Science (AAAS). 1993. *Benchmarks for science literacy:* New York: Oxford University Press.

Johmann, C. A., and E. Rieth. 1999. *Bridges: Amazing structures to design, build, and test.* Charlotte, VT: Williamson Publishing Co.

National Research Council (NRC). 1996. *National science education standards.* Washington, DC: National Academy Press.

Royce, C. A. 2007. If you build It. *Science and Children* 44 (7): 14–15.

Simon, S., and N. Fauteux. 2003. *Let's try it out with towers and bridges.* New York: Simon and Schuster.

Wolfe, M. F. 2000. *Rube Goldberg: Inventions*! New York: Simon and Shuster.

CHAPTER 9
COOL IT, DUDE!

Rosa and Paula stopped in at their favorite sub shop for a sub and a cold drink. As Rosa filled her cup with ice she complained to her friend John, the counter boy. "Why do you guys only give us crushed ice in our drinks instead of ice cubes? The ice melts so fast, we get watered down drinks instead of the real thing."

John rolled his eyes. If he had a quarter for every time that question was asked, he could quit this job and retire.

"The boss says crushed ice cools the drinks off faster," John mumbled.

"What did you say, John boy?" teased Rosa.

"I said," John said slowly and clearly, "the boss says crushed ice cools the drink quicker than cubes."

"Is that really true?" asked Rosa. "Sounds weird. Ice is ice, cubes or crushed. What difference does the size of the ice make?"

Paula piped in, "Well, I do believe crushed ice melts faster, but it waters down the drink faster too!"

"Yeah but people usually drink it fast when they eat

subs so it doesn't take them long to empty the cup," replied John. "They want their drink to be cold, fast!"

"I'm not sure I believe that stuff about cooling drinks off faster with crushed ice. You want to prove that to me?" said Paula.

"I don't have to," said John. "I only work here. Find out one way or another for yourself."

"Okay we will," said Rosa, "as soon as we get home."

"And another thing, why use crushed ice? Cause it takes up more space and we get less drink?"

John couldn't let that one go by. "It doesn't make any difference, crushed or cubes, takes up the same space.

"Wait a sec," Rosa responded, " you mean to tell me, if I crushed a cube of ice, it would take up the same space as the cube, uncrushed?"

"Well, I think so," answered John. "At least that's what the boss says. Actually, it doesn't really make a lot of sense, come to think of it. Seems like there *is* more of it, crushed."

The three looked at each other for a moment. Paula finally said, "It has to take up more space, there are more pieces."

John thought a moment. "Well, it's still the same cube—just smaller pieces— we didn't add anything."

"Yeah, but each tiny piece takes up space and there are more ice pieces to fill the cup," countered Rosa.

"Well, it looks like you have two things to prove, now " said John as he wiped off the counter. "Be sure and let me know so I can clue the boss in and get fired!" he said with a grin.

PURPOSE

There are two concepts at work here: conservation of matter and the question about many surfaces vs. fewer surfaces absorbing heat. You may wonder what this story is doing in the Earth system science area, but it has to do not only with thermodynamics and conservation of matter but with water, arguably the most important material we have on the planet. You may think it sounds more like a physics story, but. I suspect that more integration of the various sciences is done in Earth science than in any other of the disciplines because it uses so many concepts from other areas in order to understand its overarching view of the world we live in. I also feel that we do too much compartmentalizing of disciplines so that students often do not see how they all fit together. In any case, Rosa and Paula shouldn't have trouble finding out some answers to their dilemmas in their own kitchens, and nor should your students.

RELATED CONCEPTS

- Conservation of substance
- Displacement
- Heat absorption and surface area
- State changes
- Water
- Solids and liquids

DON'T BE SURPRISED

There are several possible things your students may have already formed misconceptions about that you might want to think about as you use this story. Young children may believe that if you break something into pieces, there is more of it than when it was whole. Students usually understand the flaw in this kind of thinking on their own when they are ready, but helping them see that the mass of things does not change when their shape is modified or broken into pieces is not time wasted.

The other misconception may exist into adulthood and may be semantic as much as it is scientific: Many people believe that cold moves out of ice cubes into the drink and thus cools it. This is contrary to the scientific view that *thermal energy* moves from the warmer to the cooler. In fact, *heat* is defined as a transfer of energy from an object that is hot to one that is cooler. You may want to suggest that they think of thermal energy as the mover and of cold as a lack of thermal energy. It is the warmth that does the moving, not the cold. I will address this in more detail in the content background section. It is imperative that they think of energy transfer in this way if they are to make sense out of the story and the cooling of a drink by use of ice.

CONTENT BACKGROUND

Have you ever seen children break cookies into pieces so they have more to eat? Well, they may do this until they have developed beyond what Piaget calls the *preoperational* stage of development. Then they realize that if more was not added or taken away, the amount is the same, whole or broken. This is important in this story if one is comparing the mass of an ice cube as a whole to the mass of the ice cube after it has been crushed. If your students take two identical ice cubes, crush one and leave the other whole, their masses will remain the same. They can verify this on a balance or by melting the two setups and comparing the amounts of water in each. Students will need to be satisfied that both ice cubes are the same size to begin with.

But, if your students wonder if a cupful of crushed ice by volume is the same as a cupful of whole ice cubes, it turns out to be a different story. Think of it this way: because of the shape of the cube, a cupful will have a great deal of air space between cubes. With crushed ice, the particles are smaller and fit more closely together leaving little space between each piece of ice. It seems logical that if the cup sizes are equal, more crushed ice will fit into a cup than ice cubes. One way to show this is to fill a cup with crushed ice and another cup of the same size with cubes. Allow them to melt and you will find more water in the crushed ice cup than the cube cup. So, it seems that if you are dispensing a drink and fill your cup with crushed ice, you will have less room for the drink than if you filled your cup with cubes.

The question then becomes, how much crushed ice do you put into your cup so that you are not cheating yourself on the amount of drink you purchase? This can only be decided by how fast you wish to cool your drink and how fast you drink it. There is no formula, because all ice crushers are different and therefore the particle size is different. The best answer is probably that since ice is free, you can always go back to the dispenser and replenish your ice but not your drink. So if you put a small amount of crushed ice into the cup, and your drink is not cold enough after a few minutes, you can always add more ice.

So, about the other question in the story—will crushed ice cool the drink faster? Heat transfer takes place at the interface of the surfaces of the cooler object. Let's say that we have an ice cube that is 3 cm^2. Since there are six sides to the cube, there is approximately 54 (3 cm × 3 cm × 6 sides) square centimeters of surface area touching the liquid. If you were to cut up the same full-sized cube into 1-cm cubes, there would be 27 little cubes, each with a surface of 1 cm^2. Note that there is still the same amount of ice but it is now broken up into smaller pieces with more surface area to interact with the liquid. So 27 (cubes) × 6 (sides) × 1 gives you 162 cm^2 of surface area, or three times that of a whole cube. It is no wonder that with the additional surface area interacting with the liquid, heat would be absorbed more quickly. But, with the quicker cooling the ice is melting more quickly, diluting the drink as Rosa complained. It is a dilemma! For the math and the idea for showing the increase in surface area I thank the Worsley School in Alberta, Canada, and their website, which is full of great ideas for teachers on all subjects: *www.worsleyschool.net/science/sciencepg.html.* Note that it is true the ice crusher does not crush the ice into nice little cubes, but it does create more particles with more surface area to interact with the drink and absorb the heat.

Heat and temperature are two entirely different things. Heat is commonly referred to as the amount of *energy* in a substance, and the measure of this for

everyday purposes is by a thermometer, which determines the average amount of heat in a substance or body. Temperature is a human-devised concept set up on arbitrary scales such as Fahrenheit, Celsius, or Kelvin.

Every substance has some heat in it unless it has somehow miraculously reached the temperature of absolute zero, a temperature almost impossible to attain even in a laboratory. Absolute zero is reached when no more heat energy can be extracted from a substance. The larger the substance, the more heat is present. Two ice cubes have twice the heat energy as one ice cube. A seemingly puzzling fact is that the water in an almost frozen swimming pool at 3°C has more heat in it than a glass of water at a temperature of 100° C merely because there is a great deal more water in the pool, and the amount of potential and kinetic energy in any substance is directly related to the amount of the substance.

Heat energy is attributed to the motion of atoms in any substance. More atomic activity means greater heat and less activity means less heat. So, when you heat or cool something, you are changing the activity level of its atoms. Also, heat energy can be transferred from one substance to another. By adding energy, the amount of heat a thing contains increases. A heat donor, such as the Sun, electricity, burner, or nearby higher energy source transfers its energy to the heat receiver. This is essentially the First Law of Thermodynamics.

Heat transfer, from the warmer to the cooler, can be done by one of three methods: *conduction, radiation,* or *convection.* You have felt the result of *conduction* when you put a spoon into a hot cup of liquid and then touched the spoon. The heat energy is transferred directly from the collision of the atoms in the liquid to the atoms in the spoon to you. You may also have felt the transfer of energy by *radiation* if you stood close to a fire, an electric heater, or a lamp. The energy of the heat source is in the form of infrared energy (a part of the light spectrum), which in turn excites your heat sensors and you feel heat. In *convection,* the atoms in a liquid or gas set up a current of rising and falling atoms that eventually bring everything in the substance to the same temperature. An interesting phenomenon about conduction is that some substances conduct heat better than others. For instance, if you touch metal, it feels cooler than other substances in a room. This is because the heat from your body transfers more quickly to the metal and it feels cooler to you. If the metal has been in the room for a long time, it will have the same temperature as the rest of the objects in the room. Your body will be fooled into thinking that the metal is cooler when it is really not.

Many students live with a common misconception that cold is a form of energy that can move from one place to another. They may also reckon that there is an unlimited supply of this "cold" in the ice that can continue to move into the drink and drop the temperature until the ice is gone. In their minds, the "cold" in the ice disappears into the drink until it is all used up. If one believes this, it is entirely possible for the drink to become colder than the temperature of the ice itself. We know this to be untrue since the heat of the drink will cause the ice to melt rather than allowing the ice to continue to decrease the temperature of the drink.

This phenomenon can be tested with the aid of a thermometer and a glass of ice water. The heat in the drink changes the phase of the ice from solid to liquid by increasing the energy in the atoms in the ice. This will eventually result in temperature equilibrium between the ice and the drink. When equilibrium has been reached, the temperature will go no lower because there can be no further flow of energy since everything is the same.

With crushed ice, the heat transfer will be faster because of the amount of surface area, but, again, once equilibrium is reached the drink will become no cooler. Because the ice is crushed and has more surface area, this will occur more quickly than it would with whole ice cubes.

Your students should have a great time arguing this one out and in the process they will be dealing with the properties of water in its various states and with the laws of thermodynamics as applied to the cold drink dispensers so prevalent in their daily lives.

related ideas from National Science education standards (NrC 1996)

K–4: Properties of Objects and Materials

- Materials can exist in different states—solid, liquid and gas. Heating or cooling can change some common materials, such as water, from one state to another.

K–4: Light, Heat, Electricity and Magnetism

- Heat can be produced in many ways such as burning, rubbing, or mixing one substance with another. Heat can move from one object to another.

5–8: Transfer of Energy

- Energy is a property of many substances and is associated with heat, light, electricity, mechanical motion, sound, nuclear energy and the nature of a chemical change. Energy is transferred in many ways.
- Heat moves in predictable ways, flowing from warmer objects to cooler ones until both reach the same temperature.

related ideas from BENCHMARKS FOR SCIENCE LITERACY (aaas 1993)

K–2 *The Structure of Matter*
- Heating and cooling cause changes in the properties of materials. Many kinds of changes occur faster under hotter conditions.

3–5 *Energy Transformations*
- When warmer things are put with cooler ones, the warm ones lose heat and the cool ones gain it until they are all the same temperature. A warmer object can warm a cooler one by contact or at a distance.

6–8 *Energy Transformations*
- Heat can be transferred through materials by the collision of atoms or across space by radiation. If the material is fluid, currents will be set up in it that aid the transfer of heat.
- Energy appears in different forms. Heat energy is in the disorderly motion of molecules.

USING THE STORY WITH GRADES K–4

If it makes more sense to your younger students to change the characters in the story to children and their parents, you may do so easily without changing the nature of the dilemma. True to the basic premise behind this book, this story focuses on some of the things in our lives that often go unnoticed and yet have a significant amount of science behind them.

You can get an idea of what your students are thinking about the topic by using one or both of the probes in *Uncovering Student Ideas in Science,* volumes 1 and 2 (Keeley, Eberle, and Farrin 2005; Keeley, Eberle, and Tugel 2007). "Ice Cubes in a Bag" is in volume 1 and "Ice Cold Lemonade" is in volume 2. In "Ice Cubes in a Bag," students are asked to decide if ice gains or loses mass when it melts. In "Ice Cold Lemonade," they are asked what they think about "coldness" and "heat." Both probes will evoke discussion and allow you to decide if they are ready to discuss the story. "Ice Cubes in a Bag" is best done with the ice in a bag as a prop and putting it on a balance to see if anything changes as the ice melts. Of course it doesn't. This allows you to try the investigation in several ways if the students are not convinced. They might want to use more ice cubes or just repeat the investigation.

The "Ice Cold Lemonade" probe might be difficult for younger children to fathom since the terms in the probe ask students to distinguish between heat and cold, while hot and cold are really just parts of a continuum. We often tell people to close the door to "keep the cold out," when we really should be saying, "keep the heat in." It's an example of sloppy language that helps to fuel misconceptions. For young students it is just a matter of words, but to a scientist, it is the First Law of Thermodynamics and inviolable. Heat energy moves from the warmer to the cooler. Understanding this is paramount to grasping many of the concepts revolving around energy transfer. But for young children, trying out the investigation in the story without attempting the development of the scientific concepts might be well enough in itself, and can provide supporting beams in building the scaffolding toward later understanding of energy transfer.

Third and fourth graders are very capable of reading thermometers and of setting up "fair" investigations regarding the arguments set forth in the story. It cannot be overstated that students of this age need to be reminded about conducting "fair" investigations. Variables and their control should be considered every time a new experiment is conducted. Students of this age need constant reminders about making sure that all variables except the one being tested be kept the same. Just recently I had the opportunity to interview a 10-year-old student about what a "fair" experiment meant to her. Even though she had just been reminded about variables and fairness in an interactive investigation, she replied that a "fair experiment" was one where she got the results she predicted. It is an enigma to me that some children are the first to point out unfairness in games, yet find it difficult to transfer this idea to investigations—even ones that seem similar to games. It is our role to keep reminding students about this part of research although it may seem to us repetitive and unnecessary.

Since most third and fourth graders have had the experience of using the soft drink dispensers, they are ready to test the questions asked by the story. Many will be surprised by how much difference there is in the cooling time between crushed ice and cubes and most are very surprised that so much more crushed ice fits into a drink cup. You may add an embedded assessment here and ask them how they are going to prepare their next drink at the sub shop. They can write this anecdotally in their science notebooks and this will give you a good idea about how they are applying their new knowledge.

I have had great success with an ice cube–melting race at this grade level. Students are given an ice cube and told to do anything they can to melt it except touch it with any part of their bodies. They often explain afterward that they had to get "heat" into the cube to melt it. When this happens, they are on their way to accepting that ice changes state when heat is added. I would not even attempt to distinguish between temperature and heat. Leave that for the next level.

USING THE STORY WITH GRADES 5-8

The story itself should elicit a myriad of opinions since middle schoolers have had lots of experience with soft drink machines. This can lead to consumer-based investigations on their part. If you are able to have a blender on hand to offer a ready supply of crushed ice, you and your students will have a lot of fun developing various scenarios to find the best way to get your money's worth at the dispenser. Plenty of thermometers and cups should be available. We have found that the

"micro" approach is the most economical, since you can get the same result with small cups as you can with large ones and use less ice in the process. If possible, use insulated cups so that the students' handling of their setups won't affect their results. It is a good idea to separate the two quandaries at first and have the students come up with two separate problem statements and go from there to design investigations to test their predictions. Don't forget to insist on reasons for their predictions based on some previous experience or scientific basis. This should be in their science notebooks for you to check per your scheduled timetable.

Since the National Science Education Standards suggest that trying to get your students to truly understand the difference between temperature and heat is a waste of time and effort, I suggest that you give this area minimal attention unless you are so required by your local standards. Time might be better spent having students learn to graph their data and use these data to draw conclusions and apply them to real-life situations. Does it make more sense to fill your cup with drink and then add ice until the desired temperature is reached? What kinds of problems arise with this method? Should one add ice first to a low level and then test the drink and go back for ice? What kinds of problems arise with this method? If you have access to the digital probe thermometers and the computer programs that help with the graphing and accumulating of data, your students will have even more data to display in their presentations.

I have found that if various groups come up with different tests and bring their investigation designs to the whole class to critique, everybody gets to be involved in all of the investigations to some extent. Then the groups can take the suggestions and carry out the investigations, knowing that the entire scientific community has agreed on the methodology. This helps with the sharing of data and conclusions. It can also be another embedded opportunity for formative assessment for you. Of course, this takes more time but the group construction of knowledge is well worth it.

Disagreement is inevitable and in fact should be encouraged since it is during discussions that the main conceptual understandings are built. Students should be encouraged to engage in constructive argument since this is the way that the scientific community comes to agreement. Always take this opportunity to help your students see that science is based on evidence and not opinion.

Read the article "Teaching for Conceptual Change" by Watson and Konicek (1990). This chronicles the experience of an elementary teacher whose students believe that a mitten or any wool garment produces heat. It intersperses the philosophy of inquiry teaching with the actual documentation of the teacher's attempts to allow her students to test their misconceptions against the reality of the day-to-day investigations as dictated by their questions.

related NSTA Press Books and Journal articles

Ashbrook, P. 2006. The matter of melting. *Science and children* 43 (4): 19–21.

Damonte, K. 2005. Heating up and cooling down. *Science and Children* 42 (8): 47–48.

Driver, R., A. Squires, P. Rushworth, and V. Wood-Robinson, 1994. *Making sense of secondary science: Research into children's ideas.* London and New York: Routledge Falmer.

Keeley, P., F. Eberle, and L. Farrin. 2005. *Uncovering student ideas in science: 25 formative assessment probes* (vol. 1). Arlington, VA: NSTA Press.

Keeley, P., F. Eberle, and J. Tugel. 2007. *Uncovering student ideas in science: 25 more formative assessment probes* (vol. 2). Arlington, VA: NSTA Press.

Line, L., and E. Christmann, 2004. A different phase change. *Science Scope* 28 (3): 52–53.

May, K., and M. Kurbin. 2003. To heat or not to heat. *Science Scope* 26 (5): 38.

Pusvis, D. 2006. Fun with phase changes. *Science and Children* 29 (5): 23–25.

Robertson, W. 2002. *Energy: Stop faking it! Finally understanding science so you can teach it.* Arlington, VA: NSTA Press.

references

American Association for the Advancement of Science (AAAS). 1993. *Benchmarks for science literacy.* New York: Oxford University Press.

Hazen, R., and J. Trefil. 1991. *Science Matters.* New York: Anchor Books

Keeley, P., F. Eberle, and L. Farrin. 2005. *Uncovering student ideas in science: 25 formative assessment probes* (vol. 1). Arlington, VA: NSTA Press.

Keeley, P., F. Eberle, and J. Tugel. 2007. *Uncovering student ideas in science: 25 more formative assessment probes* (vol. 2). Arlington, VA: NSTA Press.

National Research Council (NRC). 1996. National science education standards. Washington, DC: National Academy Press.

Watson, B., and R. Konicek. 1990. Teaching for conceptual change: Confronting children's Experience. *Phi Delta Kappan* 71 (9): 680–685.

Worsley School. Science and Mathematics *www.worsleyschool.net/science/sciencepg.html.*

BIOLOGICAL SCIENCES

Biological Sciences

Core concepts	Worms Are for More Than Bait	What Did That Owl Eat?	Helicopters Revisited	Flowers	A Tasteful Story
Life Cycles	X		X	X	
Classification of Organisms	X	X			
Animal Behavior	X				X
Adaptation	X	X		X	
Ecology	X	X			
Diversity of Life	X	X			
Structure and Function	X	X	X	X	X
Cells			X	X	X
Organs	X			X	X
Functions of Living Things	X	X	X	X	X
Senses		X			X
Interdependency of Living Things	X	X		X	
Needs of Organisms	X	X			
Flow of Energy	X	X			
Transformation of Matter	X	X			

CHAPTER 10

WORMS ARE FOR MORE THAN BAIT

Jim and Hal were walking to school one beautiful day in October. Jim was feeling pretty full of himself and couldn't stop talking as they walked along.

Boy, I really zinged my big sister this morning" he said.

"What did you do, put ants in her cereal?" said Hal.

"Nah, nothing as obvious as that. No, I was reading this book, *Diary of a Worm*. It's about a worm that keeps a diary, and one day he sees his sister looking in the mirror, you know, making faces and turning one way and the other. You must have seen your sister doin' that. Anyway, in the book, the worm says to his sister something like, 'No matter how much time you spend looking in the mirror, your face will always look just like your rear end.'"

Hal laughed. "For worms, I guess that's true."

"Yeah, well anyway, I said that to my sister and she freaked out and called Mom and everything. I had to apologize but it was worth it."

"Actually, how *do* you tell the front end of a worm from the back end?" asked Hal.

"I dunno. I never really tried," said Jim, after thinking a bit. "I guess, now that you bring it up, I'd kinda like to know *if* you can tell one end from the other. I never looked past putting one on my line when I go fishing."

Hal said, "I know that when my dad and I want bait for fishing, we water the lawn really wet after dark and then the earthworms come to the surface where we can catch them, but we have to use flashlights and walk very softly. They do actually come out headfirst I think, and if you're not quick enough grabbing them, they go back down the hole like lightning. I really don't know how they move back down the hole so fast, but on the ground, they can't move any faster than a slow turtle. But they sure do wiggle a lot. Come to think about it, I don't know how to tell the males from the females either."

"Let's ask Mr. Thompson if we can study them," Jim suggested. "Otherwise we may have to do something boring. Anyway, a lot of kids will be grossed out and that will be cool 'cause we won't."

They did just that and Mr. Thompson seemed delighted that the boys were interested enough in worms to want to study them. He actually was going to do something with animals anyway, so this was a perfect opportunity to engage the students. The boys already had some good questions about telling one end from the other, about distinguishing males and females, about how earthworms move, and why worms were supposed to be so good for the soil. Mr. Thompson got some live worms and the fun began.

PURPOSE

Segmented worms are considered "yucky" by a great many people, yet they are members of a large animal group that populates the entire world and provides a great service to our planet. The story should stimulate students to want to know more about this group of animals—their behavior, life cycles, habits, and their benefit to the Earth. Worms are also cool. Once kids get to know them, they find them very interesting and should become protective of these animals so vital to our ecosystem.

RELATED CONCEPTS

- Life cycles
- Classification of organisms
- Animal behavior
- Reproduction
- Adaptation
- Variation

DON'T BE SURPRISED

Worms, because of their legless bodies, their seemingly slimy coverings, and the fact that they live mostly underground, repel many students. Many people, even adults, have a number of misconceptions about worms. For instance, your students will expect that sexes are separate (i.e., males and females), not hermaphroditic. Students probably also think that you can cut worms into pieces and that all parts will produce new worms (of course, we cannot condone any experiments to investigate this). Your students may think that worms bite, yet worms have no way to do so. They may worry that worms carry disease, but worms do not. Worms are accused of eating dirt and garbage, and to this we plead on their behalf, a hearty and welcome *guilty!* Actually, it is bacteria that eat the garbage; the earthworms eat what the bacteria leave behind, as well as the bacteria themselves. Children may also believe that all bacteria are harmful. This is not true, since without bacteria decomposition would not take place. In the end, the garbage is turned into compost, a rich and nutrient-packed medium for plant growth. For that, we owe earthworms and bacteria a debt of gratitude.

CONTENT BACKGROUND

Earthworms belong to the phylum Annelida, which means "little rings," appropriately named because their bodies look like they are made up of tiny rings or segments (thus the term segmented worms). There are thousands of kinds of worms in the world, and of these, there are more than 9,000 different kinds of annelids. These may be contrasted with other kinds like the flatworms (phylum Platyhelminthes) and roundworms (phylum Nematoda). Among the three types of worms mentioned above, the majority are nonparasitic (exceptions include some species

of roundworms, which are parasitic in humans and other animals, and the tapeworm, a flatworm parasite). Earthworms range in size from half a millimeter (.02 inches) to more than 7 meters (23 feet)!

We are going to focus on the earthworm group, which includes the red worms, earthworms, and night crawlers, mainly because they are the easiest to obtain and the most common. As you may guess, earthworms live in the earth. Earthworms can be small like the red worm or large like the night crawler, which is common fishing bait. Night crawlers travel deeply into the soil, vertically down, sometimes six feet below the surface. Common earthworms tunnel more horizontally just beneath the surface. Red worms are common in vermiculture, the term for the process of turning garbage into compost. Earthworms are also a staple of the diet of predators like the mole and the shrew, which hunt them underground.

There have been entire books written on worms, several referenced at the end of the chapter, so I will only mention a few important features of earthworms that will hopefully come in handy when you have your students conduct investigations. Though they are often referred to as "lowly worms," they are quite complex physiologically. They are tubelike animals with a head and a posterior end (not really a tail). The head consists of a mouth and pharynx leading to an esophagus. At the posterior end is an anus. Inside the earthworm is a crop behind the esophagus where food is stored until it passes into the gizzard, where muscular action, along with some stones, breaks down the food into manageable pieces, after which it passes into the intestine and eventually out of the anus as a casting. In the intestine, circulating red blood (with hemoglobin) carries nutrients from the intestine into the circulatory system to nurture the earthworm's cells. The earthworm has five hearts that beat rhythmically and circulate the blood throughout the body. It has no eyes but can sense light and vibrations via nerve endings in its body. Each segment has tiny bristles or setae that are used to grasp rocks or other material in the environment so the worm can move. It does this by elongating its body, grasping the surface with its front setae and then shortening its body to catch up to the front end. The rear setae then grasp the stratum while the front end stretches out again and so on. Earthworms do not have lungs and therefore do not breathe but do respire and keep their skin moist with mucus so that they can exchange gases from the atmosphere through their skin.

Earthworms eat dirt and anything that is in that dirt and will also eat leaves and other botanical food. Thus it is a vegetarian. It is said that they eat garbage to make compost, but it is more correct to say that they eat the bacteria and fungi that decompose the garbage, which then pass through their intestine, where anything useful is extracted. Their tunneling and depositing of mineral-rich castings in the soil serves to enrich and aerate the soil, allowing it to drain better. Charles Darwin is supposed to have said that he doubted that there were any other animals that had played so important a role in the history of the world. Quite an endorsement!

Earthworms are hermaphroditic, meaning that they are bisexual. Each worm produces both eggs and sperm, and they mate by lying side by side, heads in the opposite direction in order to exchange sperm. On each worm is a saddlelike organ called a clitellum that secretes mucus, which coats the sperm and eggs and then hardens into a cocoon. Tiny worms hatch in the cocoon and emerge as tiny images of the adult.

Can you tell the front end of an earthworm from the back? You certainly can. You can infer which end is the front by allowing the worm to move and assume that most of the motion will be in a forward direction. Or you can look for the clitellum, which is located more toward the front than the rear of the body. More clinically, you can use a magnifier to look for the mouth with the overlapping prostomium on the first segment. It looks like a fat little flap of skin or lip that protects the mouth under it. Finally, the tail end is more pointed than the front of the worm.

related ideas from national science education standards (NrC 1996)

K–4: *The Characteristics of Organisms*

- Organisms have basic needs. For example animals need air, water and food;plants require air, water, nutrients and light. Organisms can survive only in environments in which their need can be met.
- The world has many different environments and distinct environments support the life of different types of organisms
- Each plant or animal has different structures that serve different functions in growth, survival, and reproduction.
- The behavior of individual organisms is influenced by external cues (such as a change in the environment). Humans and other organisms have senses that help them detect internal and external cues.

K–4: *Life Cycles of Organisms*

- Plants and animals have life cycles that include being born, developing into adults reproducing and eventually dying. The details of this life cycle are different for different organisms.
- Plants and animals closely resemble their parents.

K–4: *Organisms and Their Environments*

- All animals depend on plants. Some animals eat plants for food. Other animals eat animals that eat plants.
- An organism's patterns of behavior are related to the nature of that organism's environment, including the kinds and numbers of other organisms present, the availability of food and resources and the physical characteristics of the environment.

5–8: Structure and Function in Living Systems

- Living systems at all levels of organization demonstrate the complementary nature of structure and function. Important levels of organization for structure and function include cells, organs, tissues, organ systems, whole organisms and ecosystems.

5–8: Reproduction and Heredity

- Reproduction is a characteristic of all living systems: because no individual organism lives forever, reproduction is essential to the continuation of every species. Some organisms reproduce asexually. Other organisms reproduce sexually.

5–8: Diversity and Adaptations of Organisms

- Millions of species of animals, plants and microorganisms are alive today. Although different species might look dissimilar, the unit among organisms becomes apparent from an analysis of internal structures, the similarity of their chemical processes and the evidence of common ancestry.

RELATED IDEAS FROM BENCHMARKS FOR SCIENCE LITERACY (AAAS 1993)

K–2: Diversity of Life

- Some animals and plants are alike in the way they look and in the things they do, and others are very different from one another.
- Plants and animals have features that help them live in different environments.

3–5: Diversity of Life

- A great variety of kinds of living things can be sorted into groups in many ways using various features to decide which things belong to which group.
- Features used for grouping depend on the purpose of the grouping.

6–8: Diversity of Life

- Animals and plants have a great variety of body plans and internal structures that contribute to their being able to make or find food and reproduce.
- For sexually reproducing organisms, a species comprises all organisms that can mate with one another to produce fertile offspring.

USING THE STORY WITH GRADES K-4

This story has a lot in common with the story called "Oatmeal Bugs" from the first volume of *Everyday Science Mysteries* (Konicek-Moran 2008). For one thing, worms are about as popular with humans as beetle larvae and secondly, worms are animals that can be kept easily in a classroom or a home without worries about infestation. Worms are interesting to study because they behave in unusual ways, are not visible to the casual observer, but can perform a great service for the ecosystem.

First, ask your students to list what they already know about earthworms and place these bits of "knowledge" on chart paper. This is the "Our Best Thinking" chart that will serve as a guide throughout the study. Changing one or two words can then change these statements into questions. These questions then serve as the foci for the investigations carried out by the students.

Next is the strategy called "What Can It Tell You and What Do You Want to Know?" It begins with the students observing the worms, drawing and labeling them in their science notebooks. This is the "What Can It Tell You?" part of the strategy. It focuses on observation and not upon inference. For example, if a worm goes into the dark part of an observation chamber, a student could say, "It went into the darker side of the chamber." That is an observation. The child could also say, "The worm doesn't like the light." That would be an inference. In the first part of the strategy we focus only on observations. Inferences should have some data to back them up.

On one occasion when students were a bit sloppy about their observations, I called them to the front of the room and I drew a worm on the board as they gave me directions from their notes. As time went on, we were stymied by the lack of agreement on certain details such as the numbers of segments, the size of the clitellum, and how many setae per segment. I suggested that they needed to go back to the worm and be more careful about their observations and their notes. We returned the next day and completed the drawing successfully.

In the second part of the strategy, "What Do You Want to Know?" the students perform investigations. It should be stressed that at no time should a worm be injured or killed. One of the common misconceptions is that if you cut a worm in half, it will regenerate. Many a worm has met its demise testing this theory. The answer to the question "If you cut an earthworm in half will it make two worms?" is that it depends on what organs you cut. If the first few segments are removed, say by a bird trying to pull a worm out of its burrow, it may grow a new head. But if you cut a worm in half, the posterior end may grow another tail and then starve to death. The literature is sketchy on this whole episode of worm life, so it is best to leave it alone and ask your students to treat their worms humanely. There are plenty of other questions for your students to explore, including:

- What makes the worm back up?
- Does the worm prefer the light or the dark?
- How much does the worm eat in a day?
- What kind of temperature does the worm prefer?
- What happens when the worm meets another worm?
- What kind of soil does the worm prefer?
- Does the worm prefer a smooth surface or a rough surface?

All of these investigations can be done humanely. It's fun and instructional for the students to design the investigations and work on keeping variables controlled.

I must recommend a book to you that may help you and your students look at worms in a completely different way. It focuses on vermiculture—using worms to compost garbage. The book is called *Worms Eat Our Garbage: Classroom Activities for a Better Environment* (Appelhof, Fenton, and Harris 1993). It is full of ideas for keeping a colony of earthworms (a worm bin) in your classroom and watching garbage disappear and wonderful compost result. The format is classroom friendly with lots of activities for both younger and older students. My experience is that the most difficult aspect of studying worms is to overcome the reluctance of the teacher to have worms *in* the classroom. I must reassure you that the worms are really not a problem and if vegetable garbage is fed to the worms, there will be no odor whatsoever. If students learn to do the same at home, it could help take the strain off landfills, which would be wonderful for the environment. It demonstrates that recycling is friendly to the planet and promotes good ecological values in your students. In many areas, universities and schools are composting cafeteria garbage. With people throwing compostable food into landfills in plastic bags which will not decompose for decades, the idea of composting is an easy and valuable way for the common citizen to help the planet. Take the plunge!

USING THE STORY WITH GRADES 5–8

Many of the ideas listed above are also valid for older students. Their questions may be more sophisticated, but the worm remains the same and the investigations reward the students with valuable information. Since older students are often more familiar with thinking and planning investigations, they can get involved in more complex activities.

For example, a very simple worm observation station can be built from 12" × 18" acrylic plastic sheets and 1" × 4" boards. It is similar to ant observation structures in that it is a narrow chamber filled with dirt so that the tunnels and activities of worms below the surface can be observed. The authors of *Worms Eat Our Garbage* recommend this structure (Appelhof, Fenton, and Harris 1993). The boards are cut into 12-inch pieces to be used as the side panels; the plastic sheets are fastened to these and then attached to a base made of another board cut to 18 inches. Soil is put into the space in the middle and worms added, along with some lettuce or vegetable garbage. Paper should be placed over the plastic sheeting when the worms are not being watched since they might react to the light and burrow deeper into the chamber, making observation difficult. If possible, observations should be done under dim light, which is not as disturbing to the worms.

There are several great internet sites for more information on setting up a worm curriculum. One great example is from the University of Illinois in Urbana (*www.urbanext.uiuc.edu/worms*). The site also has a teachers' bin with lots of suggestions for activities and a question and answer section about worms in the classroom. Finally, the site shows how to use a large jar covered with paper as an alternative worm observation device, for those who are put off by constructing the one described above.

NATIONAL SCIENCE TEACHERS ASSOCIATION

If you begin with the usual chart of what children already know about worms and move to the questions that result, there should be a great number of activities that can be designed and carried out by students.

At this level, some of the questions might include

- What kinds of soil does each kind of worm prefer?
- Do different kinds of worms choose different depths at which to live?
- What kinds of food do worms dispose of most?
- Which kinds of foods are eaten most quickly?

Designing investigations for these questions can interest students and probably generate more questions. Be prepared to let the investigations go on for a long time while you plan your curriculum. Most of these investigations can be carried out at home as well, with periodic reports from the investigators from time to time.

You may well want to keep a worm bin in your classroom, as mentioned above in the K–4 section, for your students to see how well these animals recycle things that would ordinarily end up in the landfill. If your school is not already doing so, I encourage you to suggest a composting project for your school or district to your administrators. This project can prove to be a schoolwide science project and do a great deal of good for your community. Best of luck!

RELATED NSTA PRESS BOOKS AND JOURNAL ARTICLES

Driver, R., A. Squires, P. Rushworth, and V. Wood-Robinson. 1994. *Making sense of secondary science: Research into children's ideas.* London and New York: Routledge-Falmer.

Keeley, P. 2005. *Science curriculum topic study: Bridging the gap between standards and practice.* Thousand Oaks, CA: Corwin Press.

Keeley, P., F. Eberle, and L. Farrin. 2005. *Uncovering student ideas in science: 25 formative assessment probes* (vol. 1). Arlington, VA: NSTA Press.

Keeley, P., F. Eberle, and J. Tugel. 2007. *Uncovering student ideas in science: 25 more formative assessment probes* (vol. 2). Arlington, VA: NSTA Press.

Keeley, P., F. Eberle, and C. Dorsey. 2008. *Uncovering student ideas in science: Another 25 formative assessment probes* (vol. 3). Arlington, VA: NSTA Press.

REFERENCES

American Association for the Advancement of Science (AAAS). 1993. *Benchmarks for science literacy.* New York: Oxford University Press.

Appelhof, M., M. F. Fenton, and B. L. Harris. 1993. *Worms eat our garbage: Classroom activities for a better environment.* Kalamazoo, MI: Flower Press.

Cronin, D., and H. Bliss. 2003. *Diary of a worm.* New York: Scholastic Press.

Konicek-Moran, R. 2008. *Everyday Science Mysteries: Stories of inquiry-based science teaching.* Arlington, VA: NSTA Press.

National Research Council (NRC). 1996. *National science education standards.* Washington, DC: National Academy Press.

University of Illinois Extension Service Internet. Herman the Worm. *www.urbanext.uiuc.edu/worms.*

CHAPTER 11

WHAT DID THAT OWL EAT?

Enrique and Maria live in Miami, Florida, and love to visit the Everglades National Park at Shark Valley on the Tamiami Trail. It's known for its wonderful variety of water birds and especially the alligators that sun themselves alongside the 24-kilometer (15-mile) trail that winds around the sawgrass prai-rie in the Everglades. Besides the alligators, there are great blue herons, great egrets, ibis, night herons, an-hingas, cormorants, grebes, and a lot of other birds that you can see right from the path that people walk on. There is a tram that drives around the loop and an announcer gives all the details about the park and its animals and plants.

One beautiful day, Enrique and Maria were sitting under the chickee just outside the visitor's center having a sandwich and drink. A chickee is a structure made of cypress tree poles and covered with sable palm leaves, put together by workmen from the Miccosukee Indian Tribe who live nearby. The chickee is a "house without walls," a useful building for a semitropical climate like the Everglades. The chickee had benches for visitors to sit on to wait for the trams, eat lunch, or listen to programs given by the rangers and volunteers.

Because the chickee had a palm roof and cypress beams high above the benches, it was an ideal place for a barred owl to build a nest up in the rafters. On this particular day, Enrique noticed a small gray object lying on the ground under the chickee roof. He picked it up using a tissue and looked at it. It was rather solid and Enrique thought that he saw some bones and some fur in it.

Just about then, Shirley, a tram naturalist who knew just about everything about the Shark Valley area came over and asked, "What have you got there?"

"I'm not sure," said Enrique, "and neither is Maria. It was lying on the floor here and we wondered where it came from."

"I think what you have there is called an owl pellet," said Shirley. "We have a barred owl family nesting directly above you and the owls drop their pellets out of the nest onto the floor. We find one about every day."

"You mean it's owl poop?" asked Maria, who promptly dropped it onto the floor.

"No, not at all," said Shirley, "but it is a good idea that you handled it in a tissue since it might contain some germs that are not good for you. You see, owls eat their prey, usually mice or voles or even rats, whole. Inside the owl, the digestive juices dissolve the muscle parts of its prey but it doesn't digest the hair and bones of the prey."

"You mean it just sits there in its stomach?" asked Enrique.

"No," replied Shirley. "The owl's digestive system forms a mass of hair and bones that were not digested and the owl kind of spits it up and out its mouth. A lot of birds that eat their prey whole do the same thing, including grebes, cormorants, and herons."

"Yuck!" said Maria. "It's owl vomit."

"Not really," explained Shirley, "because the owl is not sick. It is just getting rid of the stuff it can't digest. Maybe you have seen your cat cough up a hair ball. It's a lot like that. In fact, if you take it apart carefully, you can find out what it has been eating. I wouldn't do that with that one if I were you since it might have some mouse germs in it but if you sterilize it for 30 minutes, it might kill the germs."

"I think I'll ask my teacher about it," said Maria. "Maybe she has some or can get some we can take apart."

And do you know what? Maria's teacher did have a way to get some owl pellets that were treated so that they were safe to take apart and see what was inside. And that is exactly what they did.

PURPOSE

This story is true, although Maria and Enrique are fictional characters. A barred owl does live in the rafter of the chickee at Shark Valley and does drop owl pellets from its nest almost daily. The purpose of this story is twofold: (1) to learn more about the eating habits of owls and (2) to learn something about the anatomy of what the owls eat, which is mostly rodents. This is done by dissecting the owl pellet and trying to put together the bones found in the pellet into a complete skeleton.

related CONCEPTS

- Skeletons
- Raptors
- Diet
- Ecology
- Predator-prey relationships

DON'T BE SURPRISED

Your students, just like Maria, may consider the pellet to be "owl poop" or "owl vomit." (Please note that scientists and naturalists refer to animal droppings as "scat.") Pellets are not scat and are common in the predator bird world as a means of getting rid of indigestible materials such as feathers, fur, and bones. Since owls eat their small prey whole, it provides us with a great opportunity to find out what they eat and to be able to put together a puzzle of bones into an entire skeleton. This is not universal in the animal world. Some predator birds, such as certain hawks, do not eat their prey whole but rip it apart and dispose of the skeleton by carrying it out of the nest. Additionally, I have dissected the scat of the ever-present alligators in Shark Valley and, to my surprise, have never found anything beyond an occasional feather. Even though the alligator eats turtles, birds, small mammals, and fish whole, the alligator's digestive system seems to absorb everything.

CONTENT BACKGROUND

Several conditions combine to cause owls to produce pellets. They eat small prey, have weak beaks for tearing their prey, and have no crop and weak acid in their digestive systems. Their food goes directly into a little pouchlike *proventriculus*, where digestion begins by means of enzymes and mucus, and then goes on to the *gizzard* since the owl has no crop for storing food. The gizzard has no enzymes or digestive juices, but grinds the prey into digestible size pieces that then pass into the intestines and are absorbed into the body. The gizzard retains the indigestible parts such as feathers, bones, and fur and compacts them into pellets that are later regurgitated to the outside. Until this happens, the owl cannot eat again, since the pellet blocks the digestive system past the gizzard.

Owls usually swoop down on their prey from a perch, using the soft feather structures of their wings to do so silently. The impact of the hit usually stuns the animal, and the owl's talons may be sufficient to kill its prey immediately. But if the quarry is larger, the owl may be forced to kill it with its beak. Small animals can be eaten immediately, but larger prey may be taken back to a perch or even stored away for leaner times. Most often owls will end up back in the nesting area, which is why pellets are found on the forest floor underneath.

Hawks, eagles, shorebirds, terns, herons, grebes, gulls, rails, shrikes, warblers, and swallows all are capable of regurgitating pellets if they eat their prey whole. These are hard to find since they do not usually regurgitate their pellets in the same place. If you find an owl pellet, it is a good idea to sterilize it by wrapping it in aluminum foil and putting it in a toaster oven at 300°F for 30 minutes before dissecting it. But the best solution is to invest in a set of owl pellets, skeleton charts, and instructions from a biological supply house. These pellets have been fumigated and should give no problem unless a student is prone to allergies. Students should still wear gloves and possibly masks. The instructions usually call for spraying the pellet with a mist of water to soften it and make it easier to tear apart.

Very often pellets from supply houses are collected in parts of the country other than that in which you teach. The origin of the pellets is usually included with the order so that you can tell where the pellets came from and possibly even what kind of owls produced them. The usual fare for owls is rats, mice, voles, and other rodents, plus an occasional earthworm or insect found near the owl's territory. We have found shrew and blackbird skeletons as well. Since the owl's digestive acid is so weak, skulls are usually present, which makes identification of the animals in the pellet easier. With older students, a magnified search of the pellet may even produce some *setae* from earthworms, but they would appear as little hairs, difficult to identify. Be sure to make it clear that you want identification sheets along with your order.

An interesting fact is that another bird group, the grebes, swallow their own feathers on a regular basis and regurgitate them up as pellets periodically. The current theory is that the feathers delay the passage of small fish bones until they can be digested or at least padded by the feathers and then later regurgitated as pellets. Since grebes live and dive in the water, it is difficult to collect their pellets for analysis. Some researchers have been able to collect some, though, and have found fish bones encased inside. I sometimes refer to the use of feathers in this way as a kind of "birdie Pepto-Bismol."

The study of pellets is both interesting and instructive. The pellet reveals what the bird has been eating and perhaps a little about its ecosystem. The bones provide an opportunity for the students to assemble animal skeletons and to learn both the names of the bones as well as their functions.

related ideas from national science education standards (nrc 1996)

K–4: The Characteristics of Organisms

- Organisms have basic needs. For example animals need air, water and food; plants require air, water, nutrients, and light. Organisms can survive only in environments in which their needs can be met.
- The world has many different environments and distinct environments support the life of different types of organisms.
- Each plant or animal has different structures that serve different functions in growth, survival, and reproduction.
- The behavior of individual organisms is influenced by external cues (such as a change in the environment).

K–4: Organisms and Their Environments

- All animals depend on plants. Some animals eat plants for food. Other animals eat animals that eat plants.
- An organism's patterns of behavior are related to the nature of that organism's environment, including the kinds and numbers of other organisms present, the availability of food and resources and the physical characteristics of the environment.

5–8: Structure and Function in Living Systems

- Living systems at all levels of organization demonstrate the complementary nature of structure and function. Important levels of organization for structure and function include cells, organs, tissues, organ systems, whole organisms, and ecosystems.

5–8: Diversity and Adaptations of Organisms

- Millions of species of animals, plants, and microorganisms are alive today. Although different species might look dissimilar, the unit among organisms becomes apparent from an analysis of internal structures, the similarity of their chemical processes and the evidence of common ancestry.

related Ideas In Benchmarks For science literacy (aaas 1993)

K–2: Diversity of Life
- Some animals and plants are alike in the way they look and in the things they do, and others are very different from one another.
- Plants and animals have features that help them live in different environments.

3–5: Diversity of Life
- A great variety of kinds of living things can be sorted into groups in many ways using various features to decide which things belong to which group.
- Features used for grouping depend on the purpose of the grouping.

6–8: Diversity of Life
- Animals and plants have a great variety of body plans and internal structures that contribute to their being able to make or find food and reproduce.

USING THE STORY WITH Grades K–4

One main deterrent to using owl pellets with younger children is the "yuck" factor. Many teachers demonstrate the dissection of the pellet before asking the children to do it. With very young children it may be enough to demonstrate the pellet and its contents as a group activity. The awe of seeing the bones in the pellet as they emerge can be as exciting to young children as it is to older children. Sometimes the students will ask to glue the pieces of the skeleton together so that it can be displayed.

Since the bones are white, it is very effective to work on a piece of black or other dark-colored paper. First, tease the pellet into halves and then into quarters so that the fur can be separated from the bones within. You can use dissecting needles or toothpicks to pull the pellet apart. Children can use sharpened pencils if you decide to let them work directly with the pellets. It usually is more effective if children work in pairs so that one can record evidence while the other dissects the pellet. They should of course trade jobs so that each gets a turn dissecting. Some children may not want to touch the bones with bare hands so rubber (nonlatex) gloves should be used. For the students who do not want to use gloves, make sure that they wash their hands after they are finished and remind them every so often to keep their hands away from their faces. No real harm would probably come from touching their faces but it is a good rule to stress whenever your students are doing lab work. An argument can be made for the use of safety glasses especially if

your school requires them. However, it is a good practice to follow for any lab and working with sharp pencils or toothpicks could accidentally cause eye damage. If any students are allergic to dust, they should also wear a nose mask.

Here one can also introduce the idea of a food chain with the children. "What did the vole eat?" (Probably an insect or a worm.) "What did the worm or insect eat?" (Probably plant material.) "Will anything eat the owl?" (Actually the owl is one of the top predators and has no natural enemies except humans, but their babies are vulnerable to foxes if they fall from the nest.) Sometimes snakes climb trees to eat babies or eggs but they risk ending up as meals themselves if the adult discovers them. Students should realize that the energy from the sun is being transferred from one organism to another as it passes up the chain since it begins with the stored energy in the bottom of all food chains, the plants.

USING THE STORY WITH GRADES 5–8

There is usually no "yuck" factor here. In fact, I was once given the advice that with middle schoolers, if all else fails, gross them out! The owl pellet is not gross but it borders on the edge enough that your students will probably take to them without hesitation. Most of the ideas described above are appropriate for the upper grades.

If your students get really interested in owls you might arrange for a guide to take them "owling." In this activity you go out about 10:00 p.m. into the woods and wait for an owl call or actual owl swooping down. Flashlights should be used at a minimum so as not to spook the owls or other animals. Actually, the best way to see any nocturnal animals is to go out about dusk and go into the woods and stand absolutely still. In no time at all, the floor of the forest will be alive with animal life. The trick is to stay still. Animals are spooked by motion, not by your presence or the color of your clothes. If you and your small group stand in a circle with everyone facing in, hands at shoulder height and watching straight ahead of them to outside the circle, you will have all points of the compass covered. Any person can signal when they see something by raising a finger and the others should slowly turn their heads to look in the direction that person indicates. Once in the woods with a group, we were visited by a doe that stood just 5 meters (15 feet) away and looked at us for five minutes before moving slowly on.

This investigation is also a wonderful opportunity to compare the skeletons of the rodents found in the pellet with the skeletal system of the human body. Many schools have a human skeleton that can be displayed along with the skeletons found in the pellet. Questions to ask are What bones do we have in common? What bones are missing or different in the rodent? In the human? What value are the bones found in each to the animal's survival?

If it is possible to obtain a feather from a hawk and one from an owl, make them available to your students for comparison. What differences do they find in them and what are the advantages of these differences? The students usually find the owl feather to be much softer and realize that the form and function match here to provide the swooping owls a silent approach to their prey. Since they are mostly nocturnal, silence is of great importance.

related NSTa Press Books and Journal articles

Driver, R., A. Squires, P. Rushworth, and V. Wood-Robinson. 1994. *Making sense of secondary science: Research into children's ideas.* London and New York: Routledge-Falmer.

Keeley, P. 2005. *Science curriculum topic study: Bridging the gap between standards and practice.* Thousand Oaks, CA: Corwin Press.

Keeley, P., F. Eberle, and L. Farrin. 2005. *Uncovering student ideas in science: 25 formative assessment probes* (vol. 1). Arlington, VA: NSTA Press.

Keeley, P., F. Eberle, and J. Tugel. 2007. *Uncovering student ideas in science: 25 more formative assessment probes* (vol. 2). Arlington, VA: NSTA Press.

Keeley, P., F. Eberle, and C. Dorsey. 2008. *Uncovering student ideas in science: Another 25 formative assessment probes* (vol. 3). Arlington, VA: NSTA Press.

Klentschy, M. 2008. *Using science notebooks in elementary classrooms.* Arlington, VA: NSTA Press.

references

American Association for the Advancement of Science (AAAS). 1993. *Benchmarks for science literacy.* New York: Oxford University Press.

National Research Council (NRC). 1996. *National science education standards.* Washington, DC: National Academy Press.

CHapter 12

TREES FROM HELICOPTERS, CONTINUED

Not quite a year had passed since Eric and Sarah had swept the maple fruits from the porch at their mother's request. They had brushed the tiny helicopters from the red maple carefully into a bag so they did not fall into the garden, germinate, and grow into little maple trees in the midst of the other flowers. Eric and Sarah were now sweeping up little red things off the porch. These were different from the year before: little, lacey, red, flowerlike objects that absolutely covered the porch floor. Eric and Sarah had not paid much attention to them previously except to moan and groan since they had a chore they did not enjoy much. Again they were told to be careful not to sweep them into the garden.

They remembered that last year they had wondered about the tiny helicopters from the maple tree and were a bit surprised when they discovered that the maple tree actually had flowers. They had found out that the helicopters were fruits of the tree that contained seeds that could germinate. They also recalled that they were amazed to find out that other trees had flowers and fruits that they had never noticed.

But these new red things were different. First of all, they did not float to the ground like helicopters, did not have wings, and were red and flimsy. No, these were definitely something else entirely.

Sarah was the first to notice that the maple tree next to the house still had many of the little red objects on them and that they were merely falling off the tree without the benefit of a wing or the wind.

"Eric, we have another mystery on our hands. These come from the same maple tree as the helicopters last year did but they are entirely different."

"Well, we still have to sweep them off so they don't turn into trees," said Eric.

"Maybe not," said Sarah. "They don't look like they are going to sprout into anything. Maybe they just make a mess and we can sweep them anywhere."

"Well, Miss Sherlock Holmes wannabe, let us go even now unto the tree and see what we can see!" Eric liked to talk like a poet sometimes just to be funny. But Sarah had heard enough of this kind of talk not to be amused.

"Okay, Eric, my boy, good idea. Let's go look at that low-hanging branch over there."

There, in miniature form were all of the data they needed in order to solve their new mystery. Neither Eric nor Sarah, who had lived next to that tree for a number of years, had ever noticed these little red things before. But whatever they were, thinking and wondering about them made the job of sweeping the porch a lot more interesting and easier for the two children, so they were grateful.

NATIONAL SCIENCE TEACHERS ASSOCIATION

PURPOSE

Maples are very interesting trees with lots of variation in form. They provide us with a view of diversity in plants as well as a chance to look at natural phenomena that is very common to anyone who has had a maple tree near them. The main purpose is to allow students to examine what happens in the reproductive life of trees. This is indeed a fascinating everyday science mystery. After all these years of sweeping our porch covered with these common objects, it was only this year that I stopped ignoring them and took the time to really observe what the tree was producing! I found I could watch the formation of the maple fruits take place day by day and begin to understand the complexities of this phenomenon. This activity also provides an excellent opportunity for students to observe, draw, and describe in their science notebooks the changes that take place over time.

RELATED CONCEPTS

- Diversity
- Reproduction
- Pollination
- Living things
- Life cycles
- Structure and function

DON'T BE SURPRISED

Unless your students have seriously been involved in an activity such as "Adopt a Tree," it is unlikely that they have noticed the complex way in which the maple tree and others of its family produce their fruits and the enclosed seeds. Many students will not be aware that trees other than fruit trees have flowers and produce fruits. Even those students who are aware of the role of flowers in the reproductive process will surely be surprised and intrigued by what they see on the maple branches in the early spring, especially examining the differences between the imperfect (male and female) flowers. You may also be taken aback to find that some of your students do not think that trees are plants.

CONTENT BACKGROUND

Maples belong to the family Aceraceae. There are about 120 species of maples in the United States. They have a very large range, found from central Florida almost to the Arctic Circle. For observational purposes, red maples (also known as swamp maples) are probably the most common east of the Mississippi, but there are also sugar maples, famous for their wonderful syrup made from the high sucrose sap of the tree in the springtime. Other maples are found in the west, such as the big leaf maple in California and the Rocky Mountain maple in the Colorado area. The box elder is in the same family and is found in all 48 states and much of Canada. All belong to the genus *Acer*; the many species will differ slightly in their leaves and flowers, but all the members of the genus will provide winged fruits.

Before the leaves come out in the spring, the flowers of the maple are very evident on the tree. First come the male flowers, which, in the case of the red maple, are bright red, with only *stamens* (the organs with pollen). Shortly after or nearly at the same time, the female flowers appear. They are on long droopy stalks that contain only *pistils* (which hold the ovaries) ready to receive the male pollen. Pollen lands on the top of the pistil, called a *stigma*. The sperm in the pollen make their way through a tube to the female ovule to fertilize it and begin the formation of the seed. Although maples are pollinated by the wind blowing the tiny pollen grains, insects that frequent both male and female flowers to drink the nectar or collect some pollen also help the process by spreading the pollen.

Plants may be either *monoecious* or *dioecious.* Monoecious organisms have both male and female reproductive parts on one plant or animal and dioecious organisms have male and female reproductive parts on separate animals or plants. My particular pet maple, the red maple, happens to be monoecious and has both male and female flowers on the same tree. However, the male flowers usually come from buds on the side of the branch and the female flowers from the buds on the tip of the twig. But this may vary from tree to tree.

Figure 12.1 Male Maple Flowers: Male *Acer rubrum*

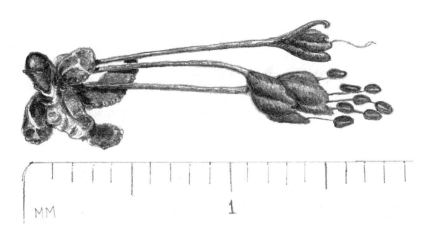

Also in my garden (and maybe in yours) is a holly bush, which is a common dioecious plant. In order for the female holly bush to produce the red berries there must be a male tree nearby. In many cities, a common dioecious tree is the *gingko*, an ancient and very beautiful tree. Most of the rest of the plants, however, are monoecious and have flowers with both male and female parts. A picture of each kind of flower is found in Figure 12.2.

Figure 12.2 Female and Male Holly Flowers: *Ilex aquilifolium*

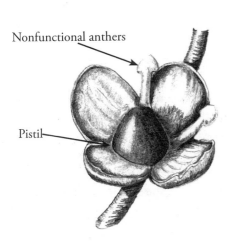

Nonfunctional anthers

Pistil

Female (Pistillate)

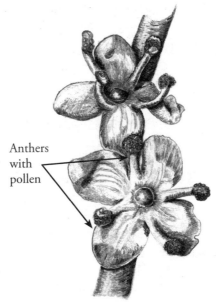

Anthers
with
pollen

Male (Staminate)

The chore given to the children was to sweep up the male flowers that fall to the ground after their time on the tree is finished. The female flowers that have been pollinated remain on the tree to produce the fruits that contain the seeds. Your students, after they have done some careful observations, will notice that the fruits are visible there, with their tiny seeds and tiny wings. The red stuff on the ground and porch Sarah and Eric had swept away are not capable of producing trees in the garden. Some of them may also be female flowers that have not been pollinated, but they are no danger either since they contain no seeds. But soon the winged fruits will be falling and the next sweeping will be a bit more difficult. But this time, they can tell Mom that they do not have to be so careful about where they sweep the flowers.

Figure 12.3 Female Maple Flowers: *Acer rubrum*

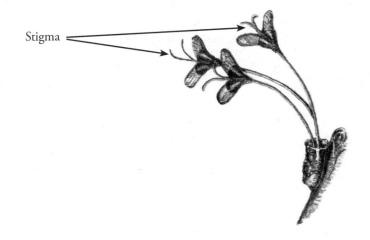

Stigma

As you can see by the drawing above, the tiny immature fruits are like miniatures of the mature fruits with which we are so familiar. For those who have not done "Trees From Helicopters" from *Everyday Science Mysteries*, it is important to note that the winged objects that fall from the trees are really fruits with seeds at the base. The ovary of the female flower has produced the seed; the wing attachment covering the seed is part of the fruit. We often refer to the winged fruits as seeds but this is not correct since the seed is only a part of the entire winged fruit. In fact, all flowering plants, called *angiosperms,* produce fruits within which lie the seeds.

On the tree, once the egg has been fertilized and the seed formed, your students will see the seed and the beginning of the wing that will function later to carry the seed away from the parent tree by wind. Fruits with wings are called *schizocarps.* If your students begin watching early enough in the process, immediately after the staminate flowers begin to drop off the tree, they will be able to witness the growth of the tiny wings into full sized wings; and they will see the stalk to which the fruit is attached elongate in order to bring the schizocarp far away from the tree branch so it can catch the wind effectively.

related Ideas From National Science education Standards (NrC 1996)

K–4: The Characteristics of Organisms

- Organisms have basic needs. For example, plants require air, water, nutrients, and light.
- Each plant or animal has different structures that serve different functions in growth, survival, and reproduction.

K–4: Life Cycles of Organisms

- Plants and animals have life cycles that include being born, developing into adults, reproducing, and eventually dying. The details of this life cycle are different for different organisms.
- Plants and animals closely resemble their parents.

6–8: Life Cycles of Organisms

- All organisms must be able to obtain and use resources, grow, reproduce and maintain stable internal conditions while living in a constantly changing external environment.

related ideas from Benchmarks for Science Literacy (aaas 1993)

K–2: The Diversity of Life

- Some animals and plants are alike in the way they look and in the things they do, and others are very different from one another.
- Plants and animals have features that help them live in different environments.

K–2: Heredity

- There is variation among individuals of one kind within a population.
- Offspring are very much, but not exactly, like their parents and like one another.

K–2: Interdependence of Life

- Living things are found almost everywhere in the world. There are somewhat different kinds in different places.

3–5: The Diversity of Life

- A great variety of kinds of living things can be sorted into groups in many ways using various features to decide which things belong to which group.
- Features used for grouping depend on the purpose of the grouping.

3–5: Heredity

- Some likenesses between children and parents such as eye color in human beings, or fruit or flower color in plants, are inherited. Other likenesses, such as people's table manners or carpentry skills, are learned.
- For offspring to resemble their parents, there must be a reliable way to transfer information from one generation to the next.

3–5: Interdependency of Life

- For any particular environment, some kinds of plants and animals survive well, some survive less well, and some cannot survive at all.
- Organisms interact with one another in various ways besides providing food. Many plants depend on animals for carrying their pollen to other plants or for dispersing their seeds.

6–8: Diversity of Life

- Animals and plants have a great variety of body plans and internal structures that contribute to their being able to make or find food and reproduce.
- For sexually reproducing organisms, a species comprises all organisms that can mate with one another to produce fertile offspring.

6–8: Heredity

- In sexual reproduction, a single specialized cell from a female merges with a specialized cell from a male. As the fertilized egg, carrying genetic information from each parent, multiplies to form the complete organism with about a trillion cells, the same genetic information is copied in each cell.

6–8: Interdependency of Life

- In all environments—freshwater, marine, forest, desert, grassland, mountain, and others—organisms with similar needs may compete with one another for resources, including food, space, water, air, and shelter. In any particular environment, growth and survival of organisms depend on the physical conditions.

USING THE STORY WITH GRADES K–4

You may like to start your lesson with a probe from *Uncovering Student Ideas in Science,* volume 2 (Keeley, Eberle, and Tugel 2007) called "Is It a Plant?" It is good to know what your students consider to be plants, and sometimes you can be surprised to find out that trees do not fit into their definition. Another possible probe is "Does It Have a Life Cycle?" from *Uncovering Student Ideas in Science,* volume 3 (Keeley, Eberle, and Dorsey 2008). Both probes will give you a formative assessment of how your students view two different concepts involving plants and life cycles. This can definitely affect the way in which you approach the topics.

I like to start with a chart where children tell me what they "know" about trees and especially about how they make seeds. This chart is called "Our Best Thinking Until Now" and can be modified as the lessons progress. Many of you may have used the old favorite exercise "Adopt a Tree" with your students. In this, students become acquainted with a tree by taking its picture over the seasons and making journal entries about changes that occur, thus observing the tree closely and getting to know it. It is a great activity because unlike other biological specimens, a tree is always where you left it, available for scrutiny. The seasonal changes make observations interesting. Patience and careful examination are necessary to get the full amount of information available.

For this particular story, it is important that you identify a maple or box elder in the general vicinity of the school or find out if your students are aware of a

maple tree in their home areas. Parental help may be useful here. Of course it is important to begin this study at the right time of the year. You can find out when the maples or elders put out flowers in your area by doing some research on the internet or asking local experts.

Once the tree or trees are found, it is important for the students to begin keeping a science notebook on this topic. They will need help in planning how they are going to observe their tree and how they are going to record data. Using the ideas in Michael Klentchy's book *Using Science Notebooks in Elementary Classrooms* (2008) would be a very helpful way to aid your students in planning what they are going to be observing. The notebook would include drawings of the flowers, labeled with dates and a method of including measurements in the drawing. With the number of digital cameras available today, photographing the flowers with a small metric ruler in the picture is an excellent method of recording data. If you desire students to draw, they can use the photograph shown on a computer as a model. My experience is that drawing pictures of objects helps students focus on details that might be missed if they merely look at a photograph. One drawing might suffice with the data for subsequent measurements entered in tabular form.

With younger students, merely drawing the flowers and fruits as they develop would be enough to help them show the differences in growth and changes in shapes and colors. When the fruits fall from the tree, those that can be germinated without a freezing period can be planted or placed in plastic bags so that the germination can be observed. See your local plant guides for requirements for germination in your section of the country. Some seeds need six months of cold in order to prepare them for germination while others, like the red maple, germinate immediately.

The intriguing thing about the development of the winged fruits is how they begin as miniature forms with the tiny wings attached to the immature fruit. The wings grow larger and larger until the fruit is ready to be disseminated. Being able to watch this daily (or, perhaps better, weekly) growth is both fascinating and informative. Maturing fruits of all sorts can, of course, be observed during the school year in climates where growing seasons last all year. In temperate climates, however, school is usually not in session during the time that squash or tomatoes, for example, ripen from the pollinated flower. Home schoolers can take advantage of this phenomenon at any time of the year due to their yearlong curriculum.

There may be a great number of "what ifs" that could be used with the fruits. For example

- What if the fruit were to be planted without the wings?
- What if the fruit were to be placed just in water without soil?
- What if the fruit were to be planted before it was ready to leave the tree on its own?
- What if the female flower had a plastic bag placed over it as soon as it developed? Would it still develop a fruit?

I am sure that your students can come up with a lot more "what ifs" and thereby develop their own inquiry-based investigations about the formation of the fruits.

USING THE STOrY WITH GraDeS 5-8

Many of the above ideas are also useful for older students. However, older students are more familiar with measurement and can carry out data gathering on their own more readily. Questions may arise on how fast the fruit wings grow or how long it takes the fruit to mature. Many questions can be asked about the flowers and pollination, such as

- What would happen if some of the female flowers were bagged and prevented from being pollinated?
- Do they notice any insects visiting the flowers and if so, what kinds of insects are doing the pollinating?
- At what point do the male flowers fall off the tree?
- Do male and female flowers appear at the same time in your tree?
- Is your tree monoecious or dioecious?
- Is there any difference between the flowers of each kind of tree (male and female)?
- Do monoecious and dioecious trees put out their flowers at the same time?

Other questions might arise with the following scenario. First, the seed is formed after pollination and fertilization, then the fruit is formed with wings and covering. Is the seed "ripe" enough to germinate before the wings are fully formed? Sounds like a research question or two in there somewhere! When you look at the young fruit developing, it looks as though the seed is fully formed, but is it? One might dissect the seeds and compare the fully formed ones with the young ones or run a germination test. The question then becomes, "Do the seeds remain on the tree only because of the need for the wings for dissemination?" I am sure that your students can find many more questions that pop into their minds about this intriguing tree. What could be the value of having two different types of flowers on the same tree or on separate trees? Some questions are good food for discussion but not necessarily for experimentation. The internet has tons of information on the particular maple or box elder that is in your location.

reLATeD NSTA Press BOOKS AND JOUrNAL ArTICLES

Driver, R., A. Squires, P. Rushworth, and V. Wood-Robinson. 1994. *Making sense of secondary science: Research into children's ideas.* London and New York: Routledge-Falmer.

Keeley, P. 2005. *Science curriculum topic study: Bridging the gap between standards and practice.* Thousand Oaks, CA: Corwin Press.

Keeley, P., F. Eberle, and L. Farrin. 2005. *Uncovering student ideas in science: 25 formative assessment probes* (vol. 1). Arlington, VA: NSTA Press.

Keeley, P., F. Eberle, and J. Tugel. 2007. *Uncovering student ideas in science: 25 more formative assessment probes* (vol. 2). Arlington, VA: NSTA Press.

Keeley, P., F. Eberle, and C. Dorsey. 2008. *Uncovering student ideas in science: Another 25 formative assessment probes* (vol. 3). Arlington, VA: NSTA Press.

references

American Association for the Advancement of Science (AAAS). 1993. *Benchmarks for science literacy.* New York: Oxford University Press.

Keeley, P. 2005. *Science curriculum topic study: Bridging the gap between standards and practice.* Thousand Oaks, CA: Corwin Press.

Keeley, P., F. Eberle, and L. Farrin. 2005. *Uncovering student ideas in science: 25 formative assessment probes* (vol. 1). Arlington, VA: NSTA Press.

Keeley, P., F. Eberle, and J. Tugel. 2007. *Uncovering student ideas in science: 25 more formative assessment probes* (vol. 2). Arlington, VA: NSTA Press.

Keeley, P., F. Eberle, and C. Dorsey. 2008. *Uncovering student ideas in science: Another 25 formative assessment probes* (vol. 3). Arlington, VA: NSTA Press.

Klentschy, M. 2008. *Using science notebooks in elementary classrooms.* Arlington, VA: NSTA Press.

Konicek-Moran, R. 2008. *Everyday Science Mysteries: Stories of inquiry-based science teaching.* Arlington, VA: NSTA Press.

National Research Council (NRC). 1996. *National science education standards.* Washington, DC: National Academy Press.

FLOWERS: MORE THAN JUST PRETTY

Olivia had a bouquet of flowers in her hand, all ready to give them to her mother for Mother's Day. There were daisies, peonies, lilies, and a few flowers she had never seen before. One thing Olivia noticed was that they were all different in one way or another. Not just in color but also in the way they were shaped and the way their little parts were arranged.

Kathleen was in the florist's shop with Olivia and was admiring the flowers too.

"Those will look very pretty on the kitchen table in a vase," said Kathleen.

"Yeah, I hope they are fresh and will last a few days," answered Olivia.

"They always seem to die so quickly when you put them in a vase of water, but outside they last for a long time."

"You know," said Kathleen, "our teacher said that everything in nature has a purpose. Now what do you suppose the purpose of flowers is? Just to look pretty when you pick them and put them in a vase?"

"It must be more than that," said Olivia. "Ms. Washington said that flowers on plants have been on Earth a lot longer than people have, so who would pick them? Chimpanzees?"

"I think chimps would be more likely to eat them than pick them for decorations," said Kathleen. "I know that my mom says that people eat some flowers, like in salads and things like that. And I think that broccoli and cauliflower are bunches of flower buds that haven't opened yet and we eat them. I think maybe asparagus, too."

"Yeah, well, maybe you eat broccoli but I avoid it whenever I can, even though everybody says it is sooooo good for me."

"Well, flowers must be good for more than food for people and bugs," said Olivia. "But why are they all so different? Like the daisies with all of those petals around the edge. They look like tiny sunflowers with the petals shining like sunlight around the middle button. And the lily with just a few petals that almost look like one big petal or like a cup full of little stalks. And they all seem to have a lot of little parts that don't make a lot of sense to me, like these things on a stalk that come out of the middle of the lily. The daisies don't have them! Or do they?"

"I don't know," said Kathleen. "I'll have to look closer. Maybe I can borrow a magnifying glass to get a better view. I think I may need a little help though since I'm not sure what all of these parts are for even after I do get a closer look at them. Maybe if we looked at them while they were still on the living plants we could find out what they do."

"Good idea," said Olivia," but I had better get these home now to Mom before they don't look as good as they do now."

NATIONAL SCIENCE TEACHERS ASSOCIATION

PURPOSE

Children love to look at flowers but few are inclined to become familiar with the structure and function of the flower. This story is aimed at providing some motivation for children to learn about one of the most important evolutionary developments in the plant and animal world. The students will also develop skills to help them decipher the purpose of these fascinating structures and the fruits that emerge from them.

RELATED CONCEPTS

- Reproduction
- Fertilization
- Pollination
- Fruits
- Seeds
- Insects

DON'T BE SURPRISED

Children may not be aware of the continuity of life in their world. They may, for example, think that seeds are dead and not a part of a continuum of flower-seed-plant-flower-seed, etc. Most are not aware of the importance of variation in the plant and animal world in developing new species or how sexual reproduction enhances the chances of new attributes being found in a population of organisms. Many will not be aware that plants can be either male or female, nor will they be aware of the role of other organisms in the cross-pollination of flowers. Working within this story framework will give you plenty of opportunity to address these concerns. These ideas are expanded and explained in the following section.

CONTENT BACKGROUND

Something amazing happened approximately 150 million years ago on Earth. Plants developed the reproductive structures we call flowers, which, in the beginning, relied on the wind to spread their pollen. In the centuries that followed, insects (probably beetles) helped flowers pollinate each other, while the plants provided food for the insects. Nice trade off! Scientists are still looking for the first flower in fossil form but as yet have not agreed on which one of the specimens found thus far is the first. About 85 million years later, butterflies, bees, wasps, and moths had evolved and joined in the flower-insect relationship. This evolutionary change in how plants reproduced altered the biological world forever.

But let's backtrack for a moment, and set the scene for the emergence of flowering plants (which are called *angiosperms*). Scientists believe that mosses were the first plants to inhabit the land about 435 million years ago, followed by ferns, ginkgos, and then by the cone-bearing plants: the *conifers*. Up to that point, the mosses and ferns produced spores that were carried either by water or wind in order to perpetuate the species. The ginkgos and conifers, being more complex, had windblown male reproductive cells that fertilized female structures, thus producing seeds not protected by fruits. These are known as the *gymnosperms* ("naked seed" in Latin). But, on the whole, pollination by these vectors was a rather hit-or-miss affair.

Then the flowering plants appeared in fossil records. Scientists are still researching about how they could have erupted so suddenly and why they came into prominence. They surmise that the introduction of flowers took sexual reproduction in plants one step beyond where it had been before, because angiosperm seeds are protected inside the female organ, the ovary. This ovary swells, produces a fruit that safeguards the seeds, and is an attractive food for animals, which help spread seeds. Angiosperm seeds were no longer destined to be naked but fully "clothed" and therefore more successful.

It is likely that the development of flowers had a strong effect upon the evolution of insects. Many plants had moved from depending upon wind to spread their pollen to depending upon insects traveling from flower to flower to spread their pollen, thereby spreading their genes beyond themselves. Flowers developed tissues that produced sugar-rich nectar at the bottom of the petals, and brilliant colors or an alluring fragrance to attract pollinators. Insects then developed a sweet tooth, a sense of smell, or color awareness, therefore finding flowers attractive places to visit. The insects get nourishment (most of the time) and the flowers get genetic diversity.

This mutual bond continues to grow even today as flowers and insects evolve to become more compatible to each other's needs. In some cases, flowers have evolved shapes that make it challenging for the insects to enter their depths, the resulting struggle through narrow chambers making it more likely that pollen will become attached. Flowers now have brighter colors and stronger fragrances. Although it is difficult to realize that the flowers are not pretty or fragrant for our delight, we are now finding many more ways the elements of their beauty attract specific kinds of insects. Bees, for example, do not see red but they do see blue and yellow. Therefore, blue and yellow bee-pollinated flowers are most likely to survive. We also know butterflies can see red, but have a poor sense of smell, so they frequent red flowers.

Bees are fond of pollen as food and collect it on their legs to transport back to the hive. As they move from flower to flower, some of the pollen is dropped or scraped off and the "goals" of the flower and the insect are realized. In some orchids, the goal is not nectar or pollen, but sex! Some flowers have evolved to look like the mating end of another bee. When the amorous bee tries to mate with the false flower "bee," the top of the flower is forced down, dabbing pollen on the back of the now frustrated bee, who moves on to another orchid to try again. Luckily for the flowers, bees learn slowly! So in essence, the insects and flowers evolve together to work toward greater success for each of the organisms involved. As a result of this fascinating process, both flowers and insects have developed what we might consider bizarre structures to accomplish the fulfillment of the needs of both parties. [David Attenborough has produced a series of films profiling the process in his series for the BBC *The Private Life of Plants.* It is available now on DVD and cassette from most distributors. This is a must-see program.

The form of flowers and the parts that make up the flower are important because of their functions. *Pollen* contains the male cells that connect with the external female structure of the flower. This is called *pollination* and is often confused with *fertilization* but is quite different. Fertilization takes place when the male reproductive cell unites with the female egg or *ovule* inside the *ovary* to form a *zygote*, which will eventually become a seed. As we noted before, insects, wind, or birds may contribute to pollination by transporting pollen to another flower of the same species. Some flowers, such as the dandelion, may self-pollinate, but usually only as a last resort.

One important thing to do is to take a look at what flowers have in common and what makes some of them unique. Flowers that rely upon wind for pollination are usually not showy since they do not have to attract insects or birds to distribute their pollen. Flowers that are showy and fragrant have adapted to attract an animal to distribute their pollen. A diagram of a simple flower is shown below and the function of each part will be described next.

Figure 13.1 Diagram of a Complete Flower

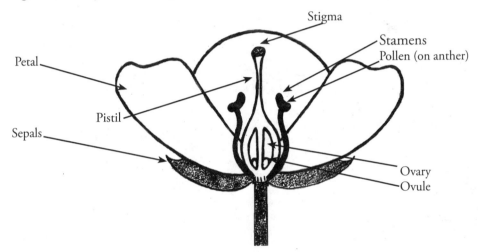

You will notice that the base of the flower (the *calyx*) is made up of *sepals*, which are not usually showy but protect the bud before the flower opens and may even fall off once the bud opens. Above the calyx is a whorl of *petals*, which are usually the pretty part of the flower, colorful in order to attract pollinators. The petals and sepals together are called the *corolla* and surround the reproductive structure(s) within. Usually in the center is the *pistil* or *carpel* (as it is often called in texts), which is comprised of the *ovary* near the base of the flower containing the *ovules* that when fertilized will become seeds. Notice the style, an erect column above the ovary and finally the *stigma*, atop the style, to which the pollen grains adhere. Ranging around the central pistil are the *stamens,* composed of the *filaments* or stalks that culminate in the *anthers* from which pollen emanates. You may notice that the anthers are lower than the pistil so self-pollination is more difficult. The main point is that trading pollen with other flowers makes genetic variation more probable and makes the production of new varieties more likely. Plants go to great lengths to avoid self-pollination. For instance, some stigmas just do not allow their own pollen to enter the pistil, or flowers ripen their sexual parts at different times.

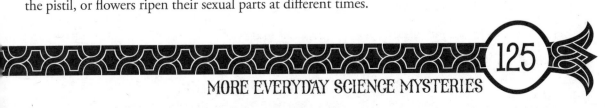

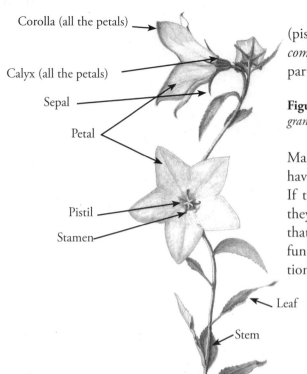

Corolla (all the petals)

Calyx (all the petals)

Sepal

Petal

Pistil

Stamen

Leaf

Stem

Flowers that have both of the reproductive parts (pistils and stamens), petals, and sepals are said to be *complete* or *perfect* flowers. If a flower has any of these parts missing it is called *incomplete* or *imperfect*.

Figure 13.2 A Complete Balloon Flower: *Platycolon grandiflora*

Male flowers are called *staminate* flowers because they have functioning stamens but no functioning pistils. If the children look carefully at, say, a maple tree, they will find *pistillate* flowers because the flowers that remain on the tree are the female flowers with functional pistils but either no stamens or nonfunctional stamens.

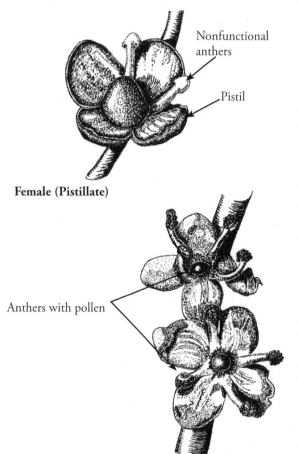

Nonfunctional anthers

Pistil

Female (Pistillate)

Anthers with pollen

Male (Staminate)

Figure 13.3 Female and Male Holly Flowers: *Ilex aquilifolium*

When pollen lands and sticks to the stigma on the pistil, it produces several nuclei within itself. One is a *tube nucleus* that bores into the style and works its way down to the ovary. The other is a male germ cell, which divides into two sperm cells that travel down the tube, where one merges with the female egg to form the *zygote* that will become the seed. The second sperm unites with other cells in the ovary to form the *endosperm*. This provides nutrition for the growing embryo. The ovary wall then develops into the fruit, which encloses the seed(s). In some plants there are multiple eggs in the ovary and multiple male cells to produce fruits with lots of seeds such as berries, corn, or peas. For an incredible set of pictures, diagrams, and explanation of this process go to *www.emc.maricopa.edu/faculty/farabee/BIOBK/BioBookflowersII.html.*

Floral reproduction is an amazing and marvelous process, besides giving us the joy of flowers' beauty. Flowers and their impact on their pollinators are vital to the ecosystems of the world. Understanding floral structure and reproduction could be the beginning of a lifelong fascination with evolution through genetic diversity.

NATIONAL SCIENCE TEACHERS ASSOCIATION

related ideas from national science education standards (Nrc 1996)

K–4: The Characteristics of Organisms

- Organisms have basic needs. For example, plants require air, water, nutrients and light.
- Each plant or animal has different structures that serve different functions in growth, survival, and reproduction.

K–4: Organisms and Their Environment

- All organisms depend on plants. Some animals eat plants for food. Other animals eat animals that eat the plants.
- An organism's patterns of behavior are related to the nature of that organism's environment, including the kinds of numbers of other organisms present, the availability of food and resources, and the physical characteristics of the environment. When the environment changes, some plants and animals survive and reproduce, and others die or move to new locations.

K–4: Life Cycles of Organisms

- Plants and animals have life cycles that include being born, developing into adults, reproducing and eventually dying. The details of this life cycle are different for different organism.
- Plants and animals closely resemble their parents.

6–8: Life Cycles of Organisms

- All organisms must be able to obtain and use resources, grow, reproduce and maintain stable internal conditions while living in a constantly changing external environment.

6–8: Structure and Function in Living Systems

- Living systems at all levels of organization demonstrate the complementary nature of structure and function. Important levels of organization for structure and function include cells, organs, tissues, organ systems, whole organisms and ecosystems.
- Specialized cells perform specialized functions in multicellular organisms. Groups of specialized cells cooperate to form a tissue, such as a muscle. Different tissues are in turn grouped together to form larger functional units, called organs. Each type of cell, tissue, and organ has a distinct structure and set of functions that serve the organism as a whole.

6–8: Reproduction and Heredity

- Reproduction is a characteristic of all living systems; because no individual organism lives forever, reproduction is essential to the continuation of every species. Some organisms reproduce asexually. Other organisms reproduce sexually.

- In many species, including humans, females produce eggs and males produce sperm. Plants also reproduce sexually—the egg and sperm are produced in the flowers of flowering plants. An egg and sperm unite to begin development of a new individual. That new individual receives genetic information from its mother (via the egg) and its father (via the sperm). Sexually produced offspring never are identical to either of their parents.

6–8: Regulation and Behavior

- An organism's behavior evolves through adaptation to its environment. How a species moves, obtains food, reproduces and responds to danger are based in the species' evolutionary history.

6–8: Diversity and Adaptations of Organisms

- Millions of species of animals, plants and microorganisms are alive today. Although different species might look dissimilar, the unity among organisms becomes apparent from an analysis of internal structures, the similarity of their chemical processes, and the evidence of common ancestry.

- Biological evolution accounts for the diversity of species developed through gradual processes over many generations. Species acquire many of their unique characteristics through biological adaptation, which involves the selection of naturally occurring variations in populations. Biological adaptations include changes in structures, behaviors or physiology that enhance survival and reproductive success in a particular environment.

related ideas from Benchmarks for science Literacy (aaas 1993)

K–2: Evolution of Life

- Different plants and animals have external features that help them thrive in different kinds of places.
- Some kinds of organisms that once lived on earth have completely disappeared, although they were something like others that are alive today.

K–2: Diversity of Life

- Some animals and plants are alike in the way they look and in the things they do, and others are very different from one another.

K–2: Heredity

- There is variation among individuals of one kind within a population.
- Offspring are very much, but not exactly, like their parents and like one another.

3–5: Diversity of Life

- A great variety of kinds of living things can be sorted into groups in many ways using various features to decide which things belong to which group.

3–5: Evolution of Life

- Individuals of the same kind differ in their characteristics, and sometimes the differences give individuals an advantage in surviving and reproducing.

3–5: Interdependence of Life

- Organisms interact with one another in various ways besides providing food. Many plants depend on animals for carrying their pollen to other plants or for dispensing their seeds.

3–5: Heredity

- Some likenesses between children and parents, such as eye color in human beings, or fruit or flower color in plants, are inherited.
- For offspring to resemble their parents, there must be a reliable way to transfer information from one generation to the next.

6–8: Diversity of Life

- Animals and plants have a great variety of body plans and internal structures that contribute to their being able to make or find food and reproduce.

6–8: Heredity

- In some kinds of organisms, all the genes come from a single parent, whereas in organisms that have sexes, typically half of the genes come from each parent.
- In sexual reproduction, a single specialized cell from a female merges with a specialized cell from a male. As the fertilized egg, carrying genetic information from each parent, multiplies to form the complete organism with about a trillion cells, the same genetic information is copied in each cell.

6–8: Evolution of Life

- Individual organisms with certain traits are more likely than others to survive and have offspring. Changes in environmental conditions can affect the survival of individual organisms and entire species.

USING THE STORIES WITH GRADES K–4

If you are interested in finding out what your students already think about plants you might want to start by giving the probe "Is It a Plant?" from *Uncovering Student Ideas in Science,* volume 2 (Keeley, Eberle, and Tugel 2007). It will give you valuable information on what kinds of organisms your students believe are plants and some insight into what they know about what attributes make something a plant. With young children you may have to read or put the plant names on a chart. Another way to do this probe with young children is to give them cards with pictures and names on them and have them sort them into "Plants" and "Not Plants" piles. Then they can tell you why they sorted the cards as they did.

After this information is in your hands, you can create a "What I Know About Flowers" chart as a "Best Thinking Until Now" activity. After the children have listed the things they "know" about flowers you can help them change their statements to questions that can be investigated. For example, children usually tell you that all flowers are pretty and smell good, or that bees and butterflies visit them. Some will say that butterflies or bees eat the flowers. At this grade level, children are seldom aware of the reproductive nature of flowers nor have they examined a flower closely.

A good place to start is with asking the children to look at flowers carefully and to draw or list things they find that they have not seen before. You might also take this opportunity to introduce your class to the use of magnifying glasses. Show them how to hold the glass up to their eyes and bring the object to be viewed up

to the glass until it is in focus. In a January 2004 article, "Discovering Flowers in a New Light," in *Science and Children,* the authors suggest the use of digital microscopes to explore the parts of a flower (McNall and Bell). The article is well worth reading. You may be fortunate enough to have such devices in your school but K–2 children are seldom adept at using them so a demonstration on a computer projection system may suffice. Young children can learn to use magnifying glasses with help, which is enough magnification to be useful in this activity.

This is also an excellent opportunity to introduce the use of science notebooks as described in Michael Klentschy's book *Using Science Notebooks in Elementary Classrooms* (2008). He suggests that the notebook be written by the students, for themselves, so that they have a mechanism to reflect on their own thinking and their own ideas.

For the initial look at plants, a flower such as the *Alstroemeria* (commonly called the Peruvian lily), usually found in the supermarket floral section, is excellent. It is large and has ample floral parts for observation even with the naked eye. It is wise to ignore the *Compositeae* flowers such as dandelions, daisies, and sunflowers, because their tiny disk in the center of the rays of petals are not easily studied and can be confusing, particularly for younger students.

I have found that observing objects for the purpose of drawing them is one of the best ways for students to focus on details. Ask the students if they have found parts of the flower they have not noticed before and if they have any idea what the parts are or what their purpose might be. I believe that it might well be enough for the students at the K–2 level to be aware of the floral parts and to find pollen on the anthers without learning the terminology of all floral parts. They should be able to learn that it is the pollen that moves with the help of insects to other flowers and has a part in developing the seeds that produce more flowers of the same kind. If your students are ready to go further, they can look at other flowers from a bouquet or their gardens and see if they can see similar parts in other flowers.

Third and fourth graders should have no trouble labeling the drawings with your help on a poster or some other diagram. But before doing so and after they have examined the flower carefully and drawn it, you may ask them if they have any ideas what the purpose of the structures might be. It would be surprising if a few students did not have some ideas to share with the class. You might focus on structure and function here, asking if they notice anything about the structures that might give them a clue as to their function. If digital microscopes are available, the pollen will be very noticeable on the anthers and perhaps even on the tip of the stigma. Good detective work on the students' part will develop some theories about what these parts might be for. They may be able to pull apart the ovaries and find the ovules, which are visible with good magnifiers but even more spectacular if you have a digital microscope. Asking them what they think these might be will get responses such as "seeds" or "pollen."

At this point, it would seem to be a good time to tell the students what botanists have discovered about the floral parts and ask the students to label their drawings using the poster or chart mentioned above. Their exploration beforehand will make your explanation even more effective because they have pondered over the mystery before you helped them to solve it. After this has been accomplished, it would be productive to ask them if they think other flowers have similar parts and to allow them to explore other flowers to see if they can compare and contrast the similarities and differences. With your knowledge from the Content Background section you should be able to help them use these comparisons and find other questions to be recorded in their notebooks for further discussion. There may be confusion if the students do try to decipher and use what they have learned while looking at flowers that do not follow exact placement of parts that they have dissected. Try to help them look at these "problem" flowers in an open way and assure them that regardless of where the parts are placed, the function and ultimate result is consistent. I think that at this age, the students should be responsible for knowing the basic principles of floral reproduction without a great deal of detail about genetics, although it is important for them to know variations in any individual can have an effect upon an entire population over time since that variation may be either successful or not and can affect an entire population. This, of course, is the basic principle of evolution.

USING THIS STORY WITH GRADES 5-8

Much of the above kind of activities can be used in the higher grades especially since students in these grades are capable of using optical or digital microscopes,— possibly more capable than we are at using the latter! I believe that dissection of the flowers is important at this level, that knowing the reproductive parts and their function will help students understand the processes that go on in the life cycle of a flowering plant. Drawing and labeling the plants in their science notebooks is essential. When students begin to ask questions about plants that are somewhat different than the simple ones they have dissected, you can help them from see that all flowers have the same function (with the exception of flowers that are either pistillate or staminate). Some flowers, particularly the *Compositeae* are more difficult to fathom because they comprise many little flowers in the center of the "flower" and usually produce multiple fruits as, for example, the dandelion and the sunflower.

If at all possible, I recommend that you work with the Wisconsin Fast Plants to allow your students to experiment with the life cycle of a plant and actually do the pollination of a flower and see the results. The fast plants are appropriately named because they go through their life cycles in approximately 40 days, seed to seed. They flower in about 14 days. You can get all the information you need to use these remarkable plants from the internet at *www.fastplants.org*. Children actually pollinate the plants by hand and can follow the life cycle of this plant right through to harvesting the next generation of seeds. The seeds and instructions can be purchased from the Carolina Biological Supply Company for between $10.00 and $25.00 depending upon how many seeds you need. You can grow them in a limited space since the plants are small and they grow under regular fluorescent light. Working with [fast plants] would be a logical extension of the explorations described above on flower anatomy and the understanding of the reproductive function of the flower. There are many investigations that can be carried out using the fast plants because the life cycle of 40 days allows for multiple experiments during a school year either for a whole class or for individuals. However, if this is not possible, watching and recording data

from flowers in your area will provide a worthwhile activity as well.

It takes a little patience, but one can watch bees and other insects or hummingbirds and observe how they feed in flowers that are compatible with their feeding style. Finding that bees and beetles actually enter the flower can be contrasted with the long proboscis feeding of the butterflies and moths and the hovering behavior of the long-beaked hummingbird and the sphinx moth. Students can find that the shape of the flower and the feeding habits of each animal are related by form and function of both feeders and plants. Hummingbirds and those insects with long beaks or long feeding tubes can suck nectar without entering the flower while the bees are required to enter into flowers, lacking extensions to their feeding mouth parts. The difference in shape of the flowers for each type of feeder should be emphasized through the concept of how form and function determine feeding habits.

Students can also watch the flowers over a period of time in a garden setting and watch what happens as the flower falls to pieces and the ovary thickens to envelope the young seeds inside. Early blooming apple blossoms in the spring produce small fruits before the end of the school year. Many early blooming flowers do the same—the process can be drawn and noted, and the small fruits of some of the plants can be dissected to find the young seeds in the fruit. Crocuses, daffodils, irises, tulips, spring beauties, and many other early bloomers produce capsules at the base of the flower so that students can watch a flower and the development of the fruit and can dissect them easily.

You could consider using two stories at once here if it is not too complicated. There is so much in common between this story and "Trees From Helicopters, Continued" that both could be used together, utilizing the flowers on the maple tree as examples of male and female flowers.

The study of flowers and their functions and their relationships with the animals with which they have entered into partnerships should provide your students with an important overview of the importance of flowering plants and their role in our local and global ecosystems.

related NSTa Press Books and Journal articles

Ashcroft, P. 2008. First explorations in flower anatomy. *Science and Children* 45 (8): 18–19.

Driver, R., A. Squires, P. Rushworth, and V. Wood-Robinson. 1994. *Making sense of secondary science: Research into children's ideas.* London and New York: Routledge-Falmer.

Keeley, P. 2005. *Science curriculum topic study: Bridging the gap between standards and practice.* Thousand Oaks, CA: Corwin Press.

Keeley, P., F. Eberle, and L. Farrin. 2005. *Uncovering student ideas in science: 25 formative assessment probes* (vol. 1). Arlington, VA: NSTA Press.

Keeley, P., F. Eberle, and J. Tugel. 2007. *Uncovering student ideas in science: 25 more formative assessment probes* (vol. 2). Arlington, VA: NSTA Press.

Keeley, P., F. Eberle, and C. Dorsey. 2008. *Uncovering student ideas in science: Another 25 formative assessment probes* (vol. 3). Arlington, VA: NSTA Press.

Klentschy, M. 2008. *Using science notebooks in elementary classrooms.* Arlington,

VA: NSTA Press.

McNall, R. L., and R. L. Bell. 2004. Discovering flowers in a new light. *Science and Children* 41 (4): 35–39.

references

American Association for the Advancement of Science (AAAS). 1993. *Benchmarks for science literacy.* New York: Oxford University Press.

Keeley, P., F. Eberle, and J. Tugel. 2007. *Uncovering student ideas in science: 25 more formative assessment probes* (vol. 2). Arlington, VA: NSTA Press.

Klentschy, M. 2008. *Using science notebooks in elementary classrooms.* Arlington, VA: NSTA Press.

McNall, R. L., and R. L. Bell. 2004. Discovering flowers in a new light. *Science and Children* 41 (4): 35–39.

National Research Council (NRC). 1996. *National science education standards.* Washington, DC: National Academy Press.

CHAPTER 14

A TASTEFUL STORY

Tyrone and Zach were spending a boring afternoon on Zach's porch. It was too hot out to do anything strenuous and since school was out for the summer they had a lot of time to do nothing. The public swimming pool was closed for the day for cleaning so they played cards for a while, and in desperation, picked up some of the magazines and newspapers that were lying around.

"Hey, Ty, look at this article about your tongue."

"What about my tongue?" said Ty.

"Not your tongue!" laughed Zach. "It's about everybody's tongue."

"You know, Zach, I know it's kinda boring this afternoon but I can't figure out why you are getting excited about an article on somebody's tongue."

"No, it's about everybody's tongue. The article in this paper Dad got at the supermarket says that we only taste certain tastes on special locations on our tongues."

"So?" said Ty impatiently.

"Well, it says that we can taste sweet on the tip of our tongues and behind the tip on both sides we can only taste salt, and behind *that* sour, and in the very back where you feel like gagging, you taste bitter things."

"So?" repeated Ty. "Makes sense to me if it's in the papers."

"You believe everything you read in papers like this one from the supermarket?"

"Well, maybe and maybe not. It depends on the paper and whether or not it makes sense."

"You know Ty, I think we ought to try this out and see if it's really true. Something doesn't seem right here because I think I can taste salt wherever it is on my tongue."

"That's easy enough to do," said Ty. "Let's go in the kitchen and find some stuff and put it on your tongue."

"Why my tongue?" said Zach. "You want to know too!"

"It's your article, so you ought to get first crack at it. Anyway, let's look it up in some books or the internet and maybe they'll tell us the truth about all of this."

So they did, and you know what? Ty and Zach got all kinds of answers and still didn't know what to believe.

"Well," said Zach, "I guess there's only one thing we can do and since we don't seem to have anything else exciting to do, I guess we ought to find out for ourselves. And I'll volunteer to go first if you will do it too. But I want to check stuff out with Mom first so we don't get into anything poison by accident."

"Okay" said Ty. "Let's see if we can get her to give us some stuff that's bitter or salty or sour or sweet that we can use."

NATIONAL SCIENCE TEACHERS ASSOCIATION

PURPOSE

Hardly a day goes by without something arriving by e-mail or being posted on the internet that just doesn't sound true. How many e-mail virus warnings have you received in the last few months? At the supermarket, we are all bombarded by news that is little more than paparazzi hype about the current celebrities or rumor, at best (not to mention stories such as those about women giving birth to two-headed alligators!). The boys in this story are depicted as having alert skepticism about things that don't actually add up in their minds and determining to find out for themselves. Actually, the tongue map has been around for a long time, even in textbooks. Just recently, it has undergone scrutiny and was found to be a myth. At least, so say some "experts." Therefore, the purpose of the story is threefold: (1) to design the proper test for the tongue map; (2) to encourage the alert skepticism we want our students to display when confronted with seemingly discrepant information; and (3) to learn something about the nervous system and how it works.

RELATED CONCEPTS

- Senses
- Propaganda
- Experimental design
- Experience
- Functions of living things
- Cells and organs
- The nervous system
- Brain function
- Urban legends

DON'T BE SURPRISED

There hasn't been a great deal of research on children's thinking about taste and, with the exception of vision, only a small amount about the other senses. However, don't be surprised if your students are unaware of the function of nerve endings on their bodies and how these send impulses to the brain where they are deciphered. In fact, many students do not understand the dominating role of the brain in everything we do, waking or sleeping. This will be especially important when you discuss with them the various taste responses—sweet, salty, bitter, and sour and possibly a new one called *umami*—which are names given to responses in our brains to certain chemicals. *Umami*, a Japanese word for "savory," is a response to glutamates found in monosodium glutamate (MSG). We should probably leave this one alone since some people are allergic to it.

Also, don't be surprised that your students might be susceptible to "urban legends" and regale you and the class with many untrue statements that have made the rounds over the years. Many of these can be debunked by quick checks in scientific references. Examples include: Gum takes seven years to digest if you swallow it; a cat always lands on its feet; warm water freezes faster than cold; lightening

doesn't strike twice in the same place; a dog's mouth is cleaner than a human's, or the yearly e-mail announcement that Mars will be as big as the Moon in the sky. There is a difference between cynicism and skepticism, and the latter will serve the student better in the long run.

CONTENT BACKGROUND

The human nervous system has specialized cells throughout the body that are connected in such a way that they communicate with once another. These cells are called *neurons*. In the tongue map, we are talking about cells that are called *receptors*. These respond to heat, pain, pressure, or, in this instance, taste. The two main nervous systems are the *central* (including the brain and the spinal cord) and the *peripheral*, which is spread all over the body and connects to the spinal cord through a set of 31 different pathways. There are three kinds of neurons in the peripheral system: the motor, the sensory, and the autonomic neurons. The autonomic neurons control your heart, lungs, digestive system, and the other parts of your body that work without your conscious attention.

We are talking about sensory neurons in this story, because they are the cells that transmit information to the central nervous system about changes within the body or outside the body. Information is sent via an electrochemical wave we call an *impulse*. The impulse travels through a part of the neuron called an *axon* to the cell's end, where it jumps the space between neurons called a *synapse*. It travels across the synapse using chemical messengers called *neurotransmitters*. The impulse thus moves on from one neuron to another until it reaches the central nervous system and the brain, where it is decoded and identified.

Sometimes certain messages are decoded by the spinal cord and the reaction is quick and unconscious. This is called a *reflex action* and happens when you touch something sharp or painful. For example, when you touch a hot stove, you withdraw your hand immediately and without thought. If you had to think about it, you might suffer damage to your body, so the body doesn't wait for you to make a decision.

The tongue is covered with tiny nerve endings called *papillae*, and in some of these papillae are the *taste buds* that transmit the nature of the chemical that has been in contact with the tongue or oral cavity. Once the message has reached the brain, it is decoded as one of the five different tastes.

At the same time, your nose is also at work through its sensory cells, and these combine with the taste transmissions to give us what we call *flavor*. Without flavor, our sense of taste is diminished and we have a difficult time telling what we are eating. You may have experienced this when you had a head cold and couldn't smell. Food doesn't taste the same. Once I was given an antibiotic that masked both my sense of smell and taste. It was very frustrating. After the antibiotic had run its course for about a month, I remember my relief as I finally was able to smell my wife's special pasta sauce cooking in the kitchen and realized that I was going to have a meal that night that I could actually taste. We take these pleasant sensations for granted until we lose them.

We have learned over time to distinguish among the many kinds of stimuli that enter our mouths. We know that most medicines and black coffee are bitter.

We know that candy is sweet and, of course, salt is salty, as are certain sauces which have been fermented, such as soy sauce. Sour can be either delightful or repulsive, depending on the strength of the acid that produces it. My grandmother used to make pickles that we called "icicle pickles" because they were so sour we shivered when eating them. Over time, I developed a craving for these extra sour pickles and found that certain brands of pickles were more sour than others.

In the early 20th century, a German scientist wrote a paper in which he mapped out certain areas of the tongue that he believed were more sensitive to certain tastes than others. In the translation of the article, the belief was probably overstated, but it caught on with the press. Even textbook authors picked it up and continued the misinformation—even to this day. This is a great lesson to all of us to be alert to written material that does not make real sense. Just because something is in print does not make it true. The best way to check it out is to try your own investigation. This experiment lends itself easily to the classroom, where the teacher can provide safe-to-use materials for the students to test on one another's tongues.

The erroneous tongue map is only one of many examples of how science information is constantly changing. For example, when I was in school studying biology, all texts said that there were 24 pairs of human chromosomes. Someone did a recount later (in Florida perhaps) and discovered that there were only 23 pairs. As science progresses, changes are made to theories in order to fit new information or to solve age-old problems. For example, the change in biological taxonomy from two kingdoms to five kingdoms occurred recently, mainly because the new classification helped to place organisms in more appropriate groups. The outdated tongue map demonstrates insufficient testing of a theory before making it into a universal fact.

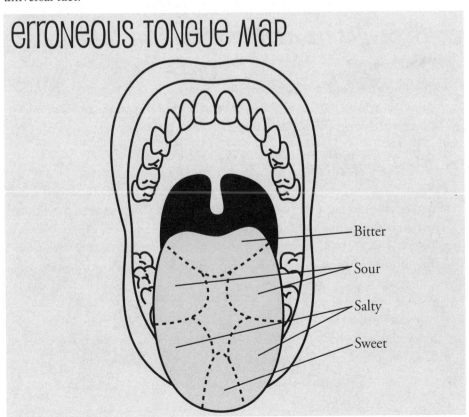

erroneous Tongue Map

Bitter

Sour

Salty

Sweet

related ideas from National science education standards (NrC 1996)

K–4: The Characteristics of Organisms

- The behavior of individual organisms is influenced by internal cues (such as hunger) and by external cues (such as a change in the environment). Humans and other organisms have senses that help them detect internal and external cues.

K–4: Understandings About Scientific Inquiry

- Scientific investigations involve asking and answering a question and comparing the answer with what scientists already know about the world.
- Scientists use different kinds of investigations depending on the questions they are trying to answer. Types of investigations include describing objects, events, and organisms; classifying them; and doing a fair test (experimenting).
- Scientists develop explanations using observations (evidence) and what they already know about the world (scientific knowledge). Good explanations are based on evidence from investigations.

5–8: Structure and Function in Living Systems

- The human organism has systems for digestion, respiration, reproduction, circulation, excretion, movement, control and coordination, and for protection from disease. These systems interact with one another.

5–8: Regulation and Behavior

- Regulation of an organism's internal environment involves sensing the internal environment and changing physiological activities to keep conditions within the range required to survive.
- Behavior is one kind of response an organism can make to an internal or environmental stimulus. A behavioral response requires coordination and communication at many levels, including cells, organ systems, and whole organisms.
- Scientists make the results of their investigations public; they describe the investigations in ways that enable others to repeat the investigations.
- Scientists review and ask questions about the results of other scientists' work.

5–8: *Abilities Necessary to Do Scientific Inquiry.*

- Identify questions that can be answered through scientific investigations. Students should be able to identify their questions with scientific ideas, concepts and quantitative relationships that guide investigations.
- Design and conduct a scientific investigation. Students can learn to formulate questions, design investigations, execute investigations, interpret data, use evidence to generate explanations, propose alternative explanations, and critique explanations and procedures.
- Think critically and logically to make the relations between evidence and explanations.
- Recognize and analyze alternative explanations and predictions.
- Understandings about scientific inquiry. Science advances through legitimate skepticism. Scientists evaluate the explanations proposed by other scientists by examining evidence. Comparing evidence, identifying faulty reasoning, pointing out statements that go beyond the evidence and suggesting alternative explanations for the same observations

related Ideas From Benchmarks For science Literacy (aaas 1993)

K–2: *The Human Organism: Basic Functions*

- The human body has parts that help it seek, find, and take in food when it feels hunger—eyes and noses for detecting food, legs to get to it, arms to carry it away and a mouth to eat it.
- Senses can warn individuals about danger; muscles help them to fight, hide, or get out of danger.
- The brain enables human beings to think and sends messages to other body parts to help them work properly.

K–2: The Scientific Enterprise

- Everybody can do science and invent things and ideas.
- In doing science, it is often helpful to work with a team and to share findings with others.

3–5: The Human Organism: Basic Functions

- The brain gets signals from all parts of the body telling what is going on there. The brain also sends signals to parts of the body to influence what they do.

3–5: The Scientific Enterprise

- Clear communication is an essential part of doing science. It enables scientists to inform others about their work, expose their ideas to criticism by other scientists, and stay informed about scientific discoveries around the world.

6–8: The Scientific Enterprise

- No matter who does science and mathematics or invents things, or when or where they do it, the knowledge and technology that result can eventually become available to everyone in the world.

The latest research seems to tell us that all of the taste buds are capable of detecting all of the tastes, even though some areas are slightly more sensitive to certain tastes than others. Most of these differences, however, are hardly measurable. I believe that if you put salt on the tip of your tongue, you will be aware of salt regardless of the "map" that says the tip of the tongue only senses sweetness.

USING THE STORY WITH GRADES K–4

After reading the story, the children will want to find out if the tongue map applies to them or not. Young children may need help understanding the function of a map. You can draw a picture of the map shown above and ask them if they believe it or not and ask how they might go about finding out for themselves. This is a good time to talk with them about variables and designing a fair test. One way to do this is to try a simple experiment with them and make some obvious unfair choices in your execution. For example, set up a ramp and place two different balls at the top of the ramp and ask which will reach the bottom first. After they have predicted, place one ball half way down the ramp or push one ball and just release the other. The cries of "not fair" will give you an opportunity to elicit from them what variables need to be controlled to have a "fair" test.

I would like to mention a word about safety here. In most classrooms, teachers forbid children to put anything into their mouths. In this activity, you have

to deviate from this rule. Assure them that the materials you have provided are completely safe and remind them that this is an exception to the rule and that they should *never* put things into their mouths unless they are given explicit instructions by you. You'll notice that in the story Zach says that he wants his mother to check the materials to make sure they are safe.

I suggest that you use either flat toothpicks or preferably cotton swabs for transferring the liquids to the tongue. Liquids are best to use, since in small amounts they cannot be inhaled by accident and the dissolved form gives better results. Lemon juice is great for sour, instant coffee is good for bitter, honey in water for sweet and, of course, saltwater for salty. You also can try to elicit from the children what they would like to try. They may have some unique ideas that would make the activity even more personal for them.

It is usually best to have the children divide into pairs and have each member of the pair test the other by placing a *small* amount (stress this) of the liquid via the cotton swab on the various sections of the tongue and see if the subject can identify the taste. The applicator should make sure all parts of the tongue map are tested. You and the students should probably devise a common data sheet for display after the activity. The sheet should include the order in which the substances were applied and the reaction of the subject to each substance. A typical response would be, "Johnny tasted salty on all parts of his tongue" or "Mary only tasted bitter in the back of her tongue." At this level it is unlikely that the students will consider that the order of tastes attempted is important, but with older age children, this might come up. Children may also like to wash out their mouths with water between tests. It might be good to provide a small container to each pair so they can spit out their rinse water, although my experience is that children don't usually need to do this.

Experience tells us that the bogus map will be disproved but some students will say that they agree with the map. You can respond that everyone is different in some little detail and that the main problem with the bogus map is that it was purported to be universal.

This lesson allows students to design and conduct an investigation with a fairly large population (the entire class). They will also have seen that all that is in print is not necessarily true.

USING THE STORY WITH GRADES 5–8

The following is an account written by a teacher, Susan Johnson, who worked with the tongue map and found that it was successful in getting children involved. I quote directly so that you can see how she handled the concepts involved and how she evaluated her own teaching as well as the students' work. Although Susan is a biology teacher, I believe that middle school students can relate to her lesson quite easily.

Traditionally, during the anatomy and physiology unit, I briefly mention the "tongue map," and show the students a diagram of this map, which divides the tongue based on salty, sweet, bitter, and sour. However, I did not allow the students the opportunity to test the tongue map; I only explained that it was developed and then recently disproved.

To make the tongue map more meaningful, I decided to modify this lesson so the students would develop the tongue map themselves. I first asked the class to develop a list of the different types of things they can taste, and they were able to come up with the four: salty, sweet, bitter, and sour. Then we brainstormed a list of foods that fit into each category. Next, the class voted on which items from their list would be best to use when developing a tongue map. They decided on the following: sweet = sugar cubes, salty = saltwater, bitter = coffee, and sour = Sour Patch Kids candy or lemon juice.

The testing material for sour became a topic for debate. Some students argued that Sour Patch Kids are sour enough that they will be sufficient for tongue mapping; however, another student pointed out that once the sour coating is gone, sour patch kids taste sweet and the sour flavor is mixed with sugar, which is sweet. They decided that lemon juice would be better for sour, since it does not have a sweet taste.

Another student noticed that three of the four taste testing items were liquid, and recommended that we use sugar water instead of sugar cubes, so that we could "swish" it around in our mouths. The class voted and decided that sugar water would be used in place of sugar cubes.

The class was then asked to develop a technique for taste testing. Students decided that they needed to rinse their mouths with fresh water between taste tests. Another student asked if the order that the different tastes are tried might affect the result. We then had to generate the number of possible taste testing orders. The students discovered 24 combinations and they wanted to try all 24. There were only 23 students in the class, so they decided that I needed to participate. Each person would get one of the combinations and would rinse with water in between each trial.

The next day I brought in lemon juice, saltwater, sugar water, and coffee. Each of us carried out our trial, and then constructed an individual tongue map on a piece of blank paper, which we then posted on the wall. The class then analyzed the various tongue maps and divided them into categories based on their similarities. Once we had grouped them, I showed them the accepted tongue map. Students who did not match the tongue map were distraught and insisted they were correct. Some students were certain they could taste all four flavors everywhere on their tongue! More investigation was needed. One student suggested we try putting each of the 4 flavors on each of the regions on our tongues.

So we did… and all the students found that they could taste all four flavors on every region of their tongues! What happened to our tongue map? Although students said that they could taste different flavors more in certain spots, they did say that they could taste all flavors over their entire tongue. Finally, I let them in on the tongue map "myth" and we read an article ("The Tongue Map: Tasteless Myth Debunked" www.livescience.com/health/060829_bad_tongue. html). This activity went well, and all of the students were really engaged. Even the students who have been very apathetic throughout the year were animated and had strong opinions when it came to arguing for their particular tongue map. Although this activity took two days to complete, I feel like it was useful. Students were able to ask questions and explore the answers to those questions.

NATIONAL SCIENCE TEACHERS ASSOCIATION

And there you have it. I could not have described the activity any better than that! Just as in any classroom, it may well be necessary to develop a common data chart so that the results can be compared. Some teachers try to wean their students from standardized data sheets.

related NSTA Press Books and Journal Articles

Driver, R., A. Squires, P. Rushworth, and V. Wood-Robinson. 1994. *Making sense of secondary science: Research into children's ideas.* London and New York: Routledge-Falmer.

Keeley, P. 2005. *Science curriculum topic study: Bridging the gap between standards and practice.* Thousand Oaks, CA: Corwin Press.

Keeley, P., F. Eberle, and L. Farrin. 2005. *Uncovering student ideas in science: 25 formative assessment probes* (vol. 1). Arlington, VA: NSTA Press.

Keeley, P., F. Eberle, and J. Tugel. 2007. *Uncovering student ideas in science: 25 more formative assessment probes* (vol. 2). Arlington, VA: NSTA Press.

Keeley, P., F. Eberle, and C. Dorsey. 2008. *Uncovering student ideas in science: Another 25 formative assessment probes* (vol. 3). Arlington, VA: NSTA Press.

Parlier, D., and M. K. Demetrikopoulos. 2004. A touch of neuroscience. *Science Scope* 28 (2): 48–50.

References

American Association for the Advancement of Science (AAAS). 1993. *Benchmarks for science literacy.* New York: Oxford University Press.

Johnson, S. Final paper for Educ. 610, University of Massachusetts. Used with the permission of the author.

National Research Council (NRC). 1996. *National science education standards.* Washington, DC: National Academy Press.

Wanjek, C. 2006. The tongue map: Tasteless myth debunked. *www.livescience.com/health/060829_bad_tongue.html.*

PHYSICAL SCIENCES

Physical Sciences

Core Concepts	The Magnet Derby	Pasta in a Hurry	Iced Tea	Color Thieves	A Mirror Big Enough
Forces	X				
Properties of Matter	X	X	X	X	
Energy	X	X	X		
Changes in State					
Heat and Temperature		X	X		
Structure of Matter			X		
Properties of Materials	X				
Light Energy				X	X
Molecules	X	X	X		
Changes of State			X		
Chemical Bonds		X	X		
Energy Spectrum				X	X
Color Spectrum				X	
Thermodynamics		X	X		
Reflection				X	X
Refraction				X	
Energy Transfer	X	X		X	X
Temperature		X			

CHAPTER 15

THE MAGNET DERBY

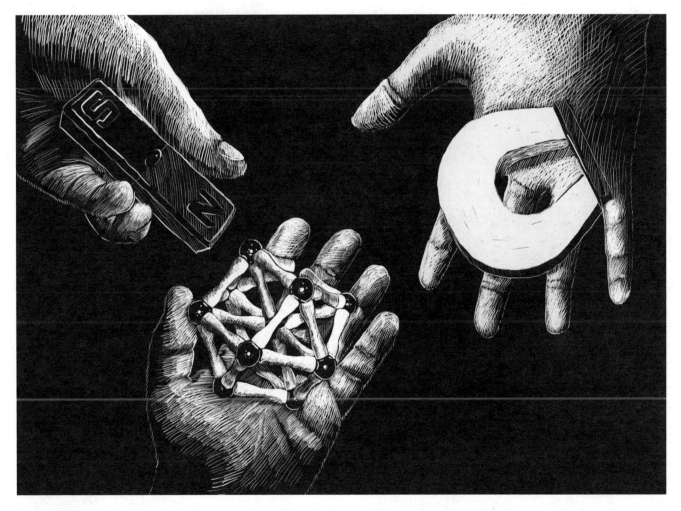

There are lots of magnets of all colors, sizes, and shapes. Some are bar magnets, some are cylinders, and others are shaped like horseshoes. Sally's teacher, Ms. Ramirez, had a whole box of them in the science cabinet.

To say that Sally liked to play with magnets was an understatement. She *loved* to play with magnets. In fact, Sally had her own collection of magnets that was bigger than the teacher's. She could never seem to get enough so when she and her dad were in the mall at the dollar store, she went in and found a group of magnets that looked more like broken pieces of metal attached to a metal cabinet. Sally tried to take one off of the cabinet and found that she could not budge it. This was the strongest magnet she had ever seen and she had to have it.

"Please Dad, this will beat all of the other magnets I have and the ones at school too, I know it!"

Sally's dad rather liked that she was fond of science things and had visions of her becoming an engineer someday.

"Okay, if you can get it off of the cabinet, you can have it," said her dad. Sally pulled and pulled but it wouldn't come straight off. Then she had an idea. The cabinet was smooth, so she thought she could slide it off the side, which she did without trouble.

"Pretty clever young lady," said her dad and together they went to the cash register.

"What makes this magnet so strong?" Sally asked the cashier.

"I'm not sure," she said, "but I think it is made of some different kinds of metals put together. They really are strong, aren't they? You should have seen us trying to get them apart when they arrived from the factory."

"Wait 'til I get this magnet to school, Dad. I'll really have the strongest magnet in the class."

Sally was disappointed that the next day was Sunday and there was no school, but on Monday morning, she ran to the bus with the magnet tucked into her pocket ready to take on all challengers. Of course, she went to Ms. Ramirez first and told her that she had a magnet that was stronger than all of the magnets in the room.

Ms. Ramirez decided she would have a little fun with Sally and also set up a neat science problem at the same time.

"Oh yeah?" asked Ms. Ramirez. "How do you think you can prove that to me and the rest of the class? We have a lot of magnets here and I'll bet some of the other kids have their favorites too. What do you say to a "Magnet Derby?"

"Sounds great to me," said Sally. "I'll go get the paper clips."

"Not so fast, Sally, I've got a different idea. Up 'til now we have always measured the strength of magnets by how many paper clips they pick up. Let's think of some different ways to measure strength so that we can have some data that is more than just numbers of clips. Anyway I'm not sure that the paper clip count is all that accurate. Remember how sometimes we got different numbers out of the same magnet? Maybe there are different ways to test strength. Let's get the class involved."

And so they did. Sally put her magnet up against all others after the class came up with several different ways of testing the strengths. They all had a lot of data at the end so that the decision was pretty certain whose magnet won the derby.

PURPOSE

Magnetism is a force that acts over a distance. Children play with magnetic toys all of the time. Any family that has a refrigerator knows that the number of magnets on the door defines the size of the appliance. This story should give students an opportunity to test some of their conceptions about magnets, namely what they will attract, their strengths, their interactions with each other, and how they are used in everyday life. It also allows for a great deal of investigation, testing of hypotheses, and drawing of conclusions.

related CONCEPTS

- Force
- Attraction
- Repulsion
- Properties of matter
- Poles

DON'T BE SURPRISED

Your students will probably expect that magnets will attract all metals. They may also believe that the strength of the magnet is determined by its size or shape. If they have been involved with measuring magnet strength before, they will assume that the number of objects they can pick up and hold judges all magnets. They will be challenged to find other ways of testing magnetic strength and to develop data collecting skills they may not have used before. They may believe that magnetism can travel through all sorts of materials. They may talk of things "sticking" to magnets rather than being attracted or pulled toward them. They may think that poles exist only at the ends of a magnet or that circular magnets have no poles. Finally, the relationship between magnets and the Earth's magnetic field will probably not occur to them. They may imagine that the magnet's needle, the north-seeking pole, is actually the north pole of the magnet.

CONTENT BACKGROUND

Magnetism is a force to be reckoned with. It is capable of doing work; it can push and pull certain objects or change their position; it can cause them to move and cause them to stop. The interesting part of the magnetic force is that it can operate through a distance—it doesn't have to touch anything. So this puts it in the same force category as gravity and electricity. In this story I am going to limit the field to permanent magnets and leave the area of electromagnets to another time. So what do we know about magnetism and magnets?

- Magnetism is a force.
- Magnetic force is created around every magnet in the form of a magnetic field.
- Magnets always have two opposite poles.

- Earth is a magnet.
- Magnetism and electricity are indelibly related.
- Like poles of magnets repel each other.
- Unlike poles of magnets attract each other.
- The forces of the magnet are strongest at the poles.
- Magnetism is a force that acts through a distance.
- Magnetic force can penetrate certain materials.

In 1861, a Scottish physicist named James Clark Maxwell said that electricity and magnetism were two aspects of the same force. He basically said that you cannot have one without the other and proved this with four mathematical equations that stand even today. He stated that every time a magnetic field changes, an electrical field is created and that every time an electric charge moves, a magnetic field is created. This means that from the magnets that adorn your refrigerator to the static electricity that gives you a bad hair day, magnetism and electricity are both involved. If you wish to delve more deeply into Maxwell's equations and this area of magnetism and electricity, I suggest you dig out your copy of *Science Matters* and read the chapter on electricity and magnetism (Hazen and Trefil 1991). I, however, shall now go on to focus on magnets as they relate to the story.

Magnetism has affected human life probably since the first human noticed the magnetic qualities of the natural magnets called lodestones. Somewhere along the line humans realized that certain metals were attracted to the lodestones and that when they were allowed to move freely, they oriented themselves in a certain direction. Later these directions were named *north* and *south*. By the same token, the end of the magnet that points in the direction we call north is called the north pole of the magnet. (Actually it was originally called the north-seeking pole and then later shortened to just north.) Its opposite was obviously named south.

Now you may well ask, if similar poles repel each other how can the north pole of a magnet or compass point to the north? Shouldn't it be repelled? Great questions, and the answers lie in the semantics of naming, maps, and directions. By convention it has been agreed that the end of the magnet that points to the pole that is in the northern hemisphere will be called north. If we think of the Earth as a magnet, then the pole that is at the top of the globe must be the south magnetic pole or else it would not attract the north pole of the compass. However, since it is in the hemisphere that we call the northern hemisphere, the pole in that hemisphere is called the north magnetic pole. It is just the opposite in the southern hemisphere. As you can see, it can seem very confusing and it's probably best not to try to introduce your students to this semantic conflict of terms. It is sufficient to concentrate on the fact that the north end of the magnet points toward the magnetic pole in the northern hemisphere and vice versa in the southern hemisphere.

In order to simplify things we paint a big *N* on one pole of the magnet and a big *S* on the other and call them north and south poles. To me, this is just another example of the fact that scientific information and nomenclature is a result of human construction. However, one interesting fact is that the poles on the Earth move around a lot, and over the years have migrated hundreds of miles. So the pole is not really at the "tip" or top of the globe we know as Earth.

Magnets can be made of nickel, cobalt, or iron. Lodestones are strictly made of iron. Magnets today are often made of combinations of the three metals. It is possible to make a temporary magnet by stroking a piece of pure iron, nickel, or cobalt with a magnet, but it will eventually lose its magnetic properties. Putting the proper metal(s) in a strong magnetic field makes permanent magnets. These do not lose their properties over time.

What kinds of things are attracted to magnets? If there is iron in something, it will be attracted to a magnet. This supposedly even includes paper if there is printing on it using ink with iron in it. Steel, which is made with iron, is attracted and, of course, the other magnetic metals such as nickel and cobalt and their alloys. And there are some rare minerals and elements, which are not likely to find their way either into our lives or our classrooms, that are attracted to magnets.

There are ceramic materials that have been infused with metals that are drawn to magnets; melding iron oxide and strontium carbonate forms one kind of ceramic magnet. All kinds of novelty displays use the flexible magnets which are made by putting ferrite magnet powder and polymers together and melding them into permanent magnets.

Magnets come in so many sizes and shapes it is difficult to predict their strength just by looking at them. There are horseshoe magnets, bar magnets, cylinder magnets, button magnets, ring magnets, and many variations of those. But they all behave alike in that they all have two poles and attract the same kinds of materials; and they all exhibit the attraction and repulsion phenomenon at the poles.

I mentioned that all magnets have two opposite poles, regardless of size. If you break a bar magnet in half, you will have two magnets, both with a north and south pole. Break it as many times as you wish and you will have smaller magnets but each will still have two poles. Even the atom is a tiny magnet with two poles. This is described by Maxwell's second equation, which says that there are no isolated poles.

Magnets are a source of fun and certainly a source of mystery, as they have been for centuries. In your class magnets should continue to stimulate your students to more and more investigations as they try to finish this story and in the meantime learn more about magnets.

related Ideas From National Science education Standards (NrC 1996)

K–4: *Light, Heat, Electricity, and Magnetism*
- Magnets attract and repel each other and certain kinds of other materials.

5–8: *Motions and Forces*
- The motion of an object can be described by its position, direction of motion, and speed. That motion can be measured and represented on a graph.

K–2: Forces of Nature

- Magnets can be used to make some things move without being touched.

3–5: Forces of Nature

- Without touching them, a magnet pulls on all things made of iron and either pushes or pulls on other magnets.
- Without touching them, material that has been electrically charged pulls on all other materials and may either push or pull other charged materials.

6–8: Forces of Nature

- Electric currents and magnets can exert a force on each other.

USING THE STORY WITH Grades K–4

Students react to this story with a competitive verve. They are ready to grab their favorite magnet and put it to the test. Again, it is important to help them define what they mean by "strongest." Since they are accustomed to using magnets to pick up things, the number of objects is usually the tack they take in defining strongest. Accept this and tell them that maybe they can find other ways to define strongest later on.

At the risk of dampening some of their competitive spirits, I suggest that you begin, as usual, with the "Best Thinking" chart and find out what they know about magnets. The chart will probably include some of the following:

- Magnets pick up things.
- Magnets only pick up metal things.
- Magnets can make things move.
- Magnets can pick up nails and pins.
- The ends of magnets either like each other or don't like each other.
- Magnets are stronger at their ends than in the middle.
- I have magnets in some of my toys.
- Magnets work under water.

Changing the first part of each statement, can change these to questions in search of evidence:

- Do magnets pick up all metal things?
- Are magnets stronger at their ends than in the middle?
- Can you stop a magnet from picking up things?

Their research and data as a class can be kept on another chart, but it is important for them to be ready to put all of their own personal results into their science notebooks (see Klentschy 2008).

Now the investigations can begin. I like to hold back some new magnets that they haven't seen so that there is an element of surprise when they design the derby later on. It is a good idea to have metal coins, aluminum foil, brass paper fasteners, paper clips, plastic bottle caps, metal bottle caps, Popsicle sticks, Legos, and lots of other objects for them to try. They will be especially surprised that some of the metal objects will not react to the magnet. You can have them sort those objects into "Don't Get Picked Up" and "Do Get Picked Up" piles. Then they can try to predict if something will be attracted to a magnet. They can use the objects that are attracted to test some of the other questions, such as on which part of the magnet the strength is greatest. After they have had a chance to really explore their magnets, they will be more ready to think about designing a magnet derby.

You may want to go to the NSTA archives and look at Peggy Ashbrook's article in *Science and Children* "More Than Messing Around with Magnets" (2005). Also useful is the article in *Science and Children* by Judith Kur and Marsha Heitzmann titled "Attracting Student Wonderings" (2008). Both articles have interesting additions to the use of magnets in K–4 classrooms.

You might give the students a hint about looking for other methods of measuring strength of magnets, such as, "You saw the paper clip move toward the magnet before it was picked up. Can you use movement of the paper clip to measure strength?" Perhaps this will allow some of your students to connect the fact that distance between magnet and object can be a measure of strength. If a paper clip is placed on lined graph paper and a magnet is brought near the clip, the number of spaces required to attract the clip to the magnet can be quantified. Object and distance can be graphed with a histogram where each magnet is measured by the number of squares required to move the clip. Students may weigh different objects and compare magnets' abilities to move them or lift them.

As a final diversion, the age-old puppet stage with cut out characters weighted down with paper clips, may allow the students to put on puppet plays by moving the characters around the stage using magnets from below.

USING THE STORY WITH GRADES 5–8

As surprising as it seems, the sophisticated middle schoolers like to play with magnets as much as the younger kids. However, we expect a bit more ingenuity from them. They should have the opportunity to put their past experience to use in new and exciting ways. Many of the ideas suggested in the last section can be modified to fit the age of the grade 5–8 students. Finding ways to measure and compare magnet strengths uses creativity and a no-holds-barred use of their minds. But once they have been able to put magnets in order from weakest to strongest, they should have an opportunity to ask questions that are seldom addressed. Unless you wish to delve into the area of electromagnets, I suggest that you ask your students to limit their investigations to the realm of permanent magnets.

Some questions to have them explore are:

- How do you limit a magnetic field?
- What kinds of materials shield a magnet's field?
- Do magnets work in a vacuum?
- Are magnets' strengths affected by a medium such as water?
- Are both poles of a particular magnet of equal strength?
- Where on a bar magnet does the polar superiority begin or end?
- Do stronger magnets also have stronger middle sections?
- Do ceramic and flexible magnets have N and S poles?

You might put questions such as these into a box and have students pick them at random and conduct investigations, then contrast and compare data with other students.

Another area that can interest your students is to take this topic into the technology realm. You might challenge them to create a levitation train such as the ones used in Germany and Japan, using only magnetic strips and everyday objects. You can make the stipulations for construction specific or leave them vague. I prefer to make certain specifications such as the materials must cost less than $5.00 and the "train" should travel in a straight path for a minimum of one meter without leaving the "track" after one push. You and the students can determine a rubric of minimum standards so that teams of students can work on the train to meet the standards.

All in all, magnets are mysterious and fun. Studying them can lead to developing all or most of the skills and knowledge specified in the Standards and Benchmarks quoted in the prior section.

RELATED NSTA PRESS BOOKS AND JOURNAL ARTICLES

Driver, R., A. Squires, P. Rushworth, and V. Wood-Robinson. 1994. *Making sense of secondary science: Research into children's ideas.* London and New York: Routledge-Falmer.

Keeley, P. 2005. *Science curriculum topic study: Bridging the gap between standards and practice.* Thousand Oaks, CA: Corwin Press.

Keeley, P., F. Eberle, and L. Farrin. 2005. *Uncovering student ideas in science: 25 formative assessment probes* (vol. 1). Arlington, VA: NSTA Press.

Keeley, P., F. Eberle, and J. Tugel. 2007. *Uncovering student ideas in science: 25 more formative assessment probes* (vol. 2). Arlington, VA: NSTA Press.

Keeley, P., F. Eberle, and C. Dorsey. 2008. *Uncovering student ideas in science: Another 25 formative assessment probes* (vol. 3). Arlington, VA: NSTA Press.

REFERENCES

American Association for the Advancement of Science (AAAS). 1993. *Benchmarks for science literacy.* New York: Oxford University Press.

Ashbrook, P. More than messing around with magnets. 2005. *Science and Children* 43 (2): 21–23.

Hazen, R., and J. Trefil. 1991. *Science matters: Achieving scientific literacy.* New York: Anchor Books.

Klentschy, M. 2008. *Using science notebooks in elementary classrooms.* Arlington, VA: NSTA Press.

Kur, J., and M. Heitzmann. 2008. Attracting student wonderings. *Science and Children* 45 (5): 24–27.

National Research Council (NRC). 1996. *National science education standards.* Washington, DC: National Academy Press.

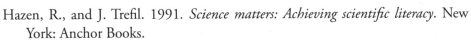

CHAPTER 16
PASTA IN A HURRY

It was almost time to leave for the movie and no one had thought about putting something on the stove for dinner. Actually it was the kids' night to cook, but they had been playing games on the computer and had forgotten the time.

Suddenly, the clock struck five times. Cavan looked up and remembered not only that she and her brother, Daniel, were supposed to make dinner that night but the movie started at 6:00 pm. Oh boy! They were in trouble and might have to skip the movie if they didn't do their job in time. There had to be time to eat and wash the dishes before they could leave. And this was a movie they had begged their parents into taking them to for weeks.

"I got it!" said Daniel, after his sister had reminded him. "We'll throw some pasta in water and open a jar of sauce. That'll be fast and get us out of this jam. Throw in a salad and we'll be eating in ten minutes."

"Sounds great," said Cavan. "Let's go. You get the water boiling for the angel hair and I'll make the salad. Angel hair is really thin so it will cook faster!"

"That sounds like a plan," said Daniel and he filled the pot with hot water to save even more time. In a few minutes, the pot was bubbling in a rolling boil and was ready for the pasta. Daniel put the pasta in and waited for the water to boil again. He wondered why putting in pasta stopped the boiling, but there was no time to worry about that now. Finally, it did boil again and Daniel turned up the heat even more.

Whoa!! The pot of pasta and water foamed up to the top of the pot. Daniel complained to Cavan that it was hard to keep the heat up high when the pasta was in it because it boiled over and made a mess.

"Turn the heat down to medium or low," said Cavan, "And it won't boil over."

"But, we're in a hurry aren't we, so I have to make the water as hot as I can!" cried Daniel.

"It's as hot as it is going to get," said Cavan. "Boiling is boiling, and I think I remember Mom telling me that once water is boiling, it doesn't get any hotter—but don't ask me why."

"No offense to Mom, but that doesn't make sense to me," said Daniel. "If I'm adding more heat to the pot the water has to get hotter! Common sense—duh!"

In the meantime, the water boiled and the pasta cooked until Daniel looked again at the pot.

"Darn, look at the water level, it's gone down, and it's not covering the pasta. I'll have to add more water and that will slow things down," complained Daniel.

"Maybe not," teased Cavan, "You put so much extra heat in there it might not slow down the boiling at all."

But it did and cooking the pasta took a little longer than Daniel had hoped.

"I know, I should have put more salt into the water so it would boil at a higher temperature."

"What does salt have to do with it?" asked Cavan.

"I'm not sure but I know that salt makes ice melt so I think it may do something to the temperature that water boils at," said Daniel.

"I don't know about all of this," said Cavan, "and the pasta looks almost done so let's let it go for tonight. We can find out later about all this heat and salt and stuff."

The meal was fine and the movie was even better than expected, but there were still some questions lingering in their heads about boiling and salt and all that other stuff.

NATIONAL SCIENCE TEACHERS ASSOCIATION

PURPOSE

This story should give students an opportunity to discover that every liquid has its own unique boiling point and that heat applied to any liquid that has reached its boiling point will not result in an increase in temperature but will be used to change the state from liquid to gas, resulting in evaporation. It also touches the point that adding substances that dissolve and ionize (break down into charged particles) to water changes the freezing and boiling points of the water.

RELATED CONCEPTS

- Energy
- Boiling and boiling points
- Change in state
- Temperature
- Heat
- Properties of matter

DON'T BE SURPRISED

The idea of a substance like water having a boiling point that will go no higher even if more heat is added is counterintuitive. As Daniel says, putting in more heat has to make it hotter. That makes sense to all of us if we look at it from our experience with lots of things in our world. In a way, Daniel is correct that heat is going into the water at a high rate but that heat is being used in a way that causes him trouble as time goes on. It speeds up evaporation—more on this in the content section. On the other side of the coin, kids know that putting salt on ice makes it melt faster, so the idea of it changing the boiling temperature isn't so far-fetched, or is it? Anyway, Daniel, Cavan, and your students have a few problems to solve about boiling points, salt, and other stuff. That "other stuff" will probably include the fact that the children, and perhaps most of the adults you know, do not realize that the bubbles you see in boiling water are bubbles of water changed to a gas and are not oxygen or any other gas.

Another misunderstanding about boiling is that some students may believe that boiling is synonymous with cooking something, regardless of the temperature. A student might say, "As long as water is boiling, shouldn't it be hot enough to cook pasta?" This may be the result of our sloppy language about boiling. We know that in a vacuum we can cause water to boil at room temperature or lower. *This* boiling water won't even feel hot. It will cook nothing; soak it, yes, cook it, no!

CONTENT BACKGROUND

To put it simply (perhaps too simply), the boiling point of any liquid is the temperature at which the liquid changes its state from liquid to vapor. It may make more sense to define it in a different way: The *boiling point* of a liquid is reached when the temperature causes the internal atmospheric pressure to equal or exceed that of the atmospheric pressure surrounding it. This means that the molecules in the liquid must have enough energy to equal or exceed the pressure surrounding the liquid, turning it into a gas. We see it as bubbles, which are, for instance in water, bubbles of water in a gaseous form that rise everywhere in the container being heated and escape from the surface of the liquid into the air as water vapor. At this point the temperature rises no more because the heat is being used to transform the liquid to gas and not to raise the temperature. The amount of heat needed to change liquid to gas is substantial.

Physical chemists and engineers often refer to vaporized water as steam. The vapor cloud seen above a teakettle or a pot is often *called* steam but it is not, since true steam is an invisible gas. What is visible is a cloud or a mist of the vapor condensing back into water droplets. In fact, if you look closely at a boiling teakettle, you will notice a space between the spout and the cloud where the steam is escaping the kettle but has not yet cooled enough to show a visible cloud.

We have direct evidence of the importance of atmospheric pressure in the boiling definition when we try to cook at higher altitudes, say in Denver or other places in altitudes of above 5,000 feet (1,500 meters). Because of the reduced atmospheric pressure at these heights, water boils at a lower temperature since the internal atmospheric pressure does not have to rise to the same temperature in order to boil, as it does at sea level. Sea level boiling point is usually listed as 212°F (100°C). Water will boil at 203° F (95° C) at an altitude of 5,000 feet (1,500 meters) and at 196° F (91° C) at 7,500 feet (2,300 meters). Even though the water is boiling, it is boiling at a lower temperature; possibly so low that *cooking* the food in the water is almost impossible. This again brings up the point that some people believe that if water is boiling, it is providing enough heat to cook. The very word "boiling" conjures up in most people's minds, HOT!! This is because we think of boiling as being a sea level phenomenon. We know that if water boils at sea level, it is dangerously hot. You would not think of plunging a hand into boiling water at sea level. Yet, in areas of very low atmospheric pressure, water may boil at so low a temperature, you would be able to submerge your hand in the water without risk of burning. We count on the water temperature being at near 100°C and so do most recipes, which give an estimated time for cooking eggs, pasta, noodles, and beans, for example. If the water is boiling at a lower temperature, the lack of heat will naturally call for an increase in the cooking time.

Some experts will tell you that beans will not cook to an edible consistency above a mile high without the use of a pressure cooker. This is a pot with a lid that increases the pressure above whatever you are cooking so that the boiling temperature is higher and therefore cooks faster. Since boiling depends on the atmospheric pressure, it is possible to boil water at room temperature or below if it is done in a vacuum. For a quick and simple look at changes of phase, I recommend reading up on it in *Science Matters* (Hazen and Trefel 1991, p. 99).

As for adding salt to the water to make it boil faster, that too would have probably been futile in a practical sense. Daniel was used to seeing his parents add salt

NATIONAL SCIENCE TEACHERS ASSOCIATION

to water when cooking grains but that would have been mostly for seasoning purposes. Some folks believe that they are raising the boiling temperature of the water, but in order to do so it would take about 2 oz (58 grams) of salt to raise 34 oz of water (one liter) 1.8°F (1°C). This amount of salt would be much too extreme for our tastes.

The scientific explanation for raising the temperature (thus the boiling point) through the application of salt is that the salt dilutes the water molecules with sodium and chlorine ions between the water molecules so their energy level is decreased. This means fewer collisions of molecules, less of a release of water vapor molecules, and therefore less pressure. Thus, a higher temperature would be needed to exceed atmospheric pressure and a higher boiling temperature would be achieved.

So you can see that Daniel is doomed to having his pasta cook at the normal rate without the use of a pressure cooker and the copious amount of salt he would have to use to hurry the process. Daniel even cost himself time by using high heat after reaching boiling temperature and causing increased evaporation of his pasta water (when it boiled over).

The above content explanation is for your benefit and perhaps not totally appropriate at this time for the students. The basic ideas to be achieved would be that boiling liquid reaches a temperature that will go no higher once boiling is achieved, that huge amounts of salt are necessary to raise the boiling point of water, and that heating water makes it evaporate faster.

related Ideas From National Science education Standards (NRC 1996)

K–4: Properties of Objects and Materials
- Materials can exist in different states—solid, liquid, and gas. Some common materials, such as water, can be changed from one state to another by heating or cooling.

5–8: Properties and Changes in Properties of Matter
- A substance has characteristic properties, such as density, a boiling point, and solubility, all of which are independent of the amount of the sample. A mixture of substances often can be separated into the original substances using one or more of the characteristic properties.

USING THE STORY WITH GRADES K–4

I assume it is obvious that young children should not be boiling water unattended because of the dangers involved. They can be asked to contribute to the "Best Thinking" chart what they "know" about boiling water. You can change these into questions, initiate the investigation, and manipulate the water and other items that the children want to explore. This does not mean that children cannot help to design an investigation. They may not be answering all the questions that Cavan and Daniel have raised, but they will have an opportunity to find out how their knowledge about boiling matches the reality of the research they create based on their own questions. For instance, it is common for very young children to want you to compare the amount of water in a pot that has been boiling for 20 minutes with the amount with which you began. Students should be encouraged to record all data in their science notebooks for further discussion.

With older children, Cavan and Daniel's questions can be addressed with the insistence that there are variables to be controlled. Again, there is no need for them to handle boiling water. A demonstration of the temperature reaching a high and then stabilizing will be just as effective done by you as if they were to do it at their own work stations. You might want to start out with a formative assessment probe to find out what your students know about boiling liquids.

Actually, the second volume of the *Understanding Student Ideas in Science* (Keeley, Eberle, and Tugel 2007) provides you with three different probes that you can use. "Turning the Dial" is aimed at finding out what students think about boiling points of water, whether they realize that boiling points are characteristics of all matter and that they remain constant regardless of how much heat is added. "Boiling Time and Temperature" elicits children's ideas about what happens to the temperature of boiling water after heat has been applied over a long time. "What's

in the Bubbles?" asks the question, "Do you know that the bubbles in boiling liquid are made up of that liquid changing state to become a gas?" Each of these can be used at different times during instruction and again at the end of the unit.

When the children have decided how to set up the investigation you may ask a student or group of students to read the thermometer and write the temperature at regular intervals on a chart or chalkboard. If you are able to borrow or if you have a probe thermometer, it will be easier to read than a regular thermometer. Be sure to measure the amount of water you start with and note it on the chart. This allows students to record all information in their own science notebooks.

As the boiling progresses, the children may notice that the amount of water is decreasing. If not, you can ask them if they think there will be as much water after the pot has boiled for 10 minutes as there was to begin with. Answers will vary but it will be necessary to let the water cool a bit and compare the before boiling amount to the after boiling amount. Do not be concerned if the temperature rises slightly after a long period of time if you are using tap water, since the impurities in the water will become more concentrated as the water evaporates and this may raise the boiling point slightly but not usually significantly. This could lead into a question about adding salt to the water to change its boiling point.

USING THE STORY WITH GRADES 5–8

You will have to use your own discretion and your school's regulations about what you allow your students to do safely in your labs. The lack of temperature change in water once it has reached boiling point will be a surprise to many students. It will bear out Cavan's statement. You may ask for predictions from your students before the investigation begins, of course asking them to give a reason for them. These should be kept in their science notebooks for future reference along with the design of the investigation and the data that results. If possible in the demonstration setup or individual setups, use a clamp or other device to suspend the thermometer or probe so that the temperature of the water is being taken, not the temperature of the container.

A spirited discussion should follow as you help them see that boiling point is a standard property of all pure matter that does not change any more than other properties. The idea of the purity of matter may raise the question brought up by Daniel about adding the salt to raise the boiling point, thus changing water from a pure substance to a "contaminated" substance.

This salt investigation will have to be designed and the amounts and increments of salt added considered by the class. My experience is that additions of five grams of salt each time the water begins to boil again will show an increase in temperature. I would like to add the caveat that you or your students put the salt added to a small amount of water in **carefully** and **slowly** so that eruptions do not occur as they sometimes do when things are added to boiling water in large amounts. Also, it should be the temperature of the water that should be measured and not of the beaker or pan, so if possible suspend the thermometer so that it does not touch the container as mentioned above. I personally would use this in demonstration form with student observers and data readers. I would also suggest that you consider using the probes mentioned in the K–4 section above.

You will find that the temperature differences will go up slowly and the question as to whether a small amount of salt added to the water would make the pasta cook any faster will be answered. It usually turns out that in order to make much of a difference, the amount of salt needed, perhaps 25 grams, would make the resulting meal very salty and not worth the small amount of time saved. There is also the point of saturation. Boiling salt water can dissolve only so much salt and if the salt added does not dissolve, it will no longer have an effect on the boiling point. Student observers may notice that after a certain amount of salt has been added, new additions will merely lie on the bottom and will not go into solution.

Students will probably have a great deal of difficulty in believing that the bubbles in the boiling liquid are water vapor. There is little you can do except to ask them to explain the decrease in volume of water in the pot as boiling continues. Invisible gases are a problem for many students and adults as well. It may be that the majority of students will understand this concept only when they are in high school and have had more experience with molecular theory. Perhaps showing students that the phase change is reversible by holding a cool plate over the boiling pot will show them that it is water in vapor form that is escaping and that it can be condensed back again into water.

I have had considerable failure in convincing *adults* about the nature of the bubbles. Their past experiences have convinced them that it is the oxygen in the water that is bubbling out. Never mind that the water seems to disappear. Certainly there is oxygen dissolved in the water, but the elimination of this amount of gas would not explain the loss of an entire pot of water if allowed to boil until gone. The misconception is well ingrained. Another possible source of confusion is that when asked what is in the space between molecules in any bit of matter, most adults will answer that it is "air." The idea of nothing but empty space between molecules is beyond their belief. This may contribute to their mistaken idea of oxygen or air bubbling out of the boiling liquid. What seems like perfect logic to us is a mystery to many students of science.

The students should now be able to finish the story about "boiling and salt and, all of that other stuff."

related NSTa Press Books and Journal articles

Damonte, K. 2005. Heating up and cooling down. *Science and Children* 42 (8): 47–48.

Driver, R., A. Squires, P. Rushworth, and V. Wood-Robinson, 1994. *Making sense of secondary science: Research into children's ideas.* London and New York: Routledge Falmer.

Keeley, P. 2005. *Science curriculum topic study: Bridging the gap between standards and practice.* Thousand Oaks, CA: Corwin Press.

Keeley, P., F. Eberle, and L. Farrin. 2005. *Uncovering student ideas in science: 25 formative assessment probes* (vol. 1). Arlington, VA: NSTA Press.

Keeley, P., F. Eberle, and J. Tugel. 2007. *Uncovering student ideas in science: 25 more formative assessment probes* (vol. 2). Arlington, VA: NSTA Press.

Keeley, P., F. Eberle, and C. Dorsey. 2008. *Uncovering student ideas in science: Another 25 formative assessment probes* (vol. 3). Arlington, VA: NSTA Press.

Klentschy, M. 2008. *Using science notebooks in elementary classrooms.* Arlington, VA: NSTA Press.

Link, L., and E. Christmann. 2004. A different phase change. *Science Scope* 28 (3): 52–54.

May, K., and M. Kurbin. 2003. To heat or not to heat. *Science Scope* 26 (5): 38.

Pusvis, D. 2006. Fun with phase changes. *Science and Children* 29 (5): 23–25.

Robertson, W. 2002. *Energy: Stop faking it! Finally understanding science so you can teach it.* Arlington, VA: NSTA Press.

references

American Association for the Advancement of Science (AAAS). 1993. *Benchmarks for science literacy.* New York: Oxford University Press.

Hazen, R., and J. Trefil. 1991. *Science matters: Achieving scientific literacy.* New York: Anchor Books.

Keeley, P., F. Eberle, and J. Tugel. 2007. *Uncovering student ideas in science: 25 more formative assessment probes* (vol. 2). Arlington, VA: NSTA Press.

National Research Council (NRC). 1996. *National science education standards.* Washington, DC: National Academy Press.

CHAPTER 17
ICED TEA

Caroline wiped the water from the outside of her glass of Kool Aid. The weather was hot and humid. The weather person said it was the hottest July on record. Caroline believed it. She was only nine years old but she couldn't remember any summer that had been so uncomfortable. The ice cubes clinked in her glass and touched her nose as she drank the deliciously cool liquid.

Lisa, her older sister, drifted out onto the porch with a tray of glasses and a pitcher of iced tea.

"I'm out on the porch!" Lisa shouted over her shoulder as she kicked the screen door shut with her left foot. That meant that the rest of the family would join them soon. The glasses rattled on the tray as she regained her balance. Lisa walked over to the table and set the tray and its contents down.

Caroline didn't move. She didn't care for tea. She had a sweet tooth and tea, even with sugar in it, didn't quite make it. Lisa was always teasing her to give up her Kool Aid, and she didn't waste any time today.

"Hey, Carrie," she said, "give up that sugar water

and have a grown-up drink. It's better for you anyway."

"I'll have a grown-up drink when I'm grown up," replied Caroline. "Right now I'm still a kid."

"What's the matter, my iced tea is not sweet enough for you?"

"I can make it as sweet as I want but I still won't like it," snapped Caroline.

"How sweet can you make it?" asked Lisa.

"I can put a whole bowl of sugar in it and then I'll have tea sugar water. So what's the difference?"

Lisa ignored her question. Instead, she retorted, "I'll bet you can't dissolve a whole bowl of sugar in a glass of this iced tea. In fact, I bet you can't even dissolve a quarter of a cup before the family gets out here."

"Bet I can," replied Caroline defiantly. "What's the bet?"

"My chores against yours for all the rest of this week."

"You're on!" said Caroline confidently.

Caroline poured herself a large glass of iced tea, ice cold right from the pitcher. She grabbed the sugar bowl and a spoon as well. Slowly she dipped the spoon into the bowl and moved a heaping spoonful to the glass of iced tea and let the sugar fall into the tea. It fell like snow, clearly visible as it trailed down to the bottom of the glass. She did this twice more until she had three teaspoonfuls of sugar in the glass of tea.

"It needs to be stirred," observed Caroline and she reached for the long handled iced teaspoon on the table. She stirred the mixture and the white stuff that no longer looked like sugar whirled up and around as she kept the liquid in motion.

She stopped stirring and watched. The cloudy, swirling substance slowly became calm and finally piled up in a sloppy mess at the bottom.

"Told you!" said Lisa, smirking.

"I'm not done yet!" said Caroline, stirring madly at the iced tea. Why won't it dissolve? she thought. Several more attempts at stirring did not produce any visible results. Well, maybe a *little*. Caroline made a move toward the sugar bowl with her spoon.

"Right!" said Lisa, sarcastically. "If it won't dissolve three teaspoons, try more."

Caroline stopped and thought. As much as she hated to admit it, her big sister had her. She decided to taste the tea. It was a little sweet, but not very sweet—not sweet enough anyway. What was the matter here? She had tasted sweetened iced tea before, yet hers was not nearly that sweet, and the sugar wouldn't dissolve. How could she get the iced tea to accept the sugar and sweeten it? There had to be a way. And why *wouldn't* the sugar dissolve?

Just at that point her thoughts were interrupted by the arrival of her family and the roll of paper towels and a bottle of spray cleaner that fell into her lap.

"I was supposed to do the kitchen windows today little sister," said Lisa. "Do a good job—remember, no streaks!"

PURPOSE

Dissolving things in the universal solvent water is an everyday experience for almost all of us. We don't have a lot of trouble dissolving honey in our hot tea, but when it comes to sweetening cold iced tea, it is quite a task, if not almost impossible. Obviously, water is not really a "universal solvent" as it is sometimes called; otherwise we couldn't find anything to keep it in. But it does dissolve more substances than any other liquid in common use. This is also a wonderful opportunity for you to show the students the effect of temperature on dissolving substances in liquid and to make the distinction between dissolving and melting. For older students, learning about the molecular structure of water might be appropriate.

RELATED CONCEPTS

- Molecules
- Solvent
- Solute
- Solution
- Melting
- Chemical bonds
- Solid
- Liquid

DON'T BE SURPRISED

One of the biggest misconceptions among children and even some adults is the distinction between dissolving and melting. Children will often say that the sugar put into a drink "melted." Melting, of course, is a change in state caused by bringing a substance to its melting point by adding heat energy. Dissolving is the combination of two or more substances into a solution. Common language and even some dictionary definitions do not help the situation since the two words are often inferred to have the same meaning. However, in science they are two completely different concepts.

CONTENT BACKGROUND

Lisa was pretty certain she would win that bet. Why? Probably because she had experienced trying to sweeten cold tea with sugar. Did you notice that she put a time limit on the bet? Eventually, with enough stirring, the sugar would probably have dissolved, but it would have taken a very long time and more patience than either Caroline or your students would have. Iced tea is usually sweetened while the tea water is still hot, which speeds up the dissolving, or it is sweetened afterward using a liquid syrup made of sugar dissolved in hot water. The liquid solution of sugar and water dissolves much faster than solid sugar in water. How is this different, then, from melting?

Raising it to its melting point by adding heat energy causes a substance to melt. Ice cream melts in warm weather, as anyone will attest if they have ever tried to finish an ice cream cone before it dribbled all over the front of his or her clothes. This is a physical change—the melted ice cream is still ice cream but in a different state. Every substance has a melting point. We don't realize this because most things we recognize as solids have melting points well above the normal temperatures we experience in our world. Perhaps some of you made lead soldiers and remembered how much heat had to be put into the solid metal in order to get it to melt so you could put it into the molds.

Solid (frozen) water, on the other hand, melts at room temperature. This shows that water is naturally in its melted state in our world except for times when the air temperature drops below its melting point. In other words, it changes to ice when the temperature around it reaches water's freezing point: 32°F or 0°C.

Water, which is used to make iced tea, is called a *solvent*, and the tea as well as the sugar are called the *solutes*. So, the dissolver is the solvent and the thing that is dissolved into the solvent is called the solute. Water dissolves more substances than any other substance on Earth. That is why it is called the "universal solvent" although, as I mentioned above, that is an exaggeration. It's a good thing, too, since it would dissolve our body's cells and everything around us since there is so much water in the world! We can't do without it for more than a day or two and it makes up more than half of every living thing.

Water is also unique in its properties (a quality of a substance which is predictable, such as the boiling or freezing point), so scientists use it as a standard for many functions. For instance, in calculating density, everything is compared to the density of water, which is 1 gram for each cubic centimeter. And, the calorie measurement in heat is based upon how much of the heat is needed to raise one cubic centimeter of water one degree Celsius.

So why is water such a good solvent? The answer lies in the shape of its molecule, which is like a tiny boomerang, and its electrical polarity. The oxygen atom is at the top or point of the boomerang and the hydrogens are at the ends of the arms. There are two hydrogens attached to one oxygen, thus the famous symbol or formula for water as H_2O. The molecule has a negative end and a positive end, thus it is called a *polar* molecule. This means that it attracts other molecules of its own kind as well as other polar molecules—of which there are plenty. The negative end attracts the positive ends of other molecules and the positive attracts the negative ends.

When another polar substance such as salt or sugar is put into water, their molecules break apart because the elements (which are now charged ions) are surrounded by the water molecules and are unable to stay attached to each other. For example, when a crystal of sugar is dropped into the water or tea, a molecule of sugar leaves the crystal and breaks apart because of the attraction of the polar molecules in the water. The elements of the crystal are immediately surrounded by the water molecules in the tea and become part of what we call a *solution*. These molecules cannot be filtered out and will not fall out due to gravity once they are in solution. If however, you should boil away all of the water, you would be left with the molecules that were dissolved as a residue on the bottom of the pan.

H_2O MOLECULE

Oxygen

Hydrogen

Any solution can be reversed and the component parts separated based on the properties of the items in the solution; for instance, water will boil away at 212° F or 100°C, but the sugar won't, which is why there is a residue. However, a filter cannot separate the various components of a solution. So there is really no chemical reaction, no change of state (in other words, no melting) and a new substance is not formed, yet the solute molecules are within the solution and are, in essence, part of the whole. The water still tastes sweet. In the process of going into solution, the solute is distributed homogeneously throughout the solution. This means that if you took a sample anywhere in the solution, the ratio of solute to solution would be the same.

Now, since the water molecules are hitting the sugar crystals at a rate commensurate with the temperature of the water, warmer water would make the water molecules more active and therefore the number of water molecules hitting the sugar crystal at any one time would be much greater. This would speed up the process of dissolving, which is why dissolving sugar in hot water is much faster than trying to dissolve it in cold tea. In fact, higher temperatures generally speed up any reaction between two or more reactants.

At some point, for reasons too complicated to explain here, a solvent can absorb no more solute. This is called the *saturation point*. However, any definition of saturation includes the qualification "at a given temperature." If the temperature is raised, the amount of solute that can go into solution increases. If the solution is then cooled, it is over-saturated and we say that the solution is *super-saturated* at that temperature.

When we think of Caroline adding more and more sugar, we know that at the ice-cold temperature of the tea, it was limited in how much sugar it could take into solution. She may have been at a saturation point or, depending upon the amount of sugar and the temperature of the tea, the sugar might have dissolved a bit more with much, much more stirring. Whatever the situation, Caroline got stuck with doing the windows and Lisa won an easy bet.

related Ideas From National Science education Standards (NRC 1996)

K–4: Properties of Objects and Materials

- Objects have many observable properties, including size, weight, shape, color, temperature, and the ability to react with other substances. These properties can be measured using tools such as rulers, balances, and thermometers.
- Materials can exist in different states—solid, liquid and gas. Some common materials such as water can be changed from one state to another by heating or cooling.

5–8: Properties and Changes of Properties in Matter

- A substance has characteristic properties, such as density, a boiling point, and solubility, all of which are independent of the amount of the sample. A mixture of substances often can be separated into the original substances using one or more of the characteristic properties.

related ideas in Benchmarks for science literacy (aaas 1993)

K–2: Structure of Matter

- Things can be done to materials to change some of their properties, but not all materials respond the same way to what is done to them.

3–5: Structure of Matter

- Heating and cooling cause changes in the properties of materials. Many kinds of changes occur faster under hotter conditions.
- Materials may be composed of parts that are too small to be seen without magnification.

6–8: Structure of Matter

- All matter is made up of atoms, which are far too small to see directly through a microscope. The atoms of any element are alike but are different from atoms of other elements. Atoms may stick together in well-defined molecules or may be packed together in large arrays. Different arrangements of atoms into groups compose all substances.
- Atoms and molecules are perpetually in motion. Increased temperature means greater average energy of motion, so most substances expand when heated.
- The temperature and acidity of a solution influence reaction rates. Many substances dissolve in water, which may greatly facilitate reactions between them.

USING THE STORY WITH Grades K–4

You may want to use the probe "Is It Melting?" from *Uncovering Student Ideas in Science,* volume 1 (Keeley, Eberle, and Farrin 2005), as a pre- or formative assessment along the way. This will help you discover what your students are thinking about the difference between dissolving and melting.

After reading the story to the children, allow them to discuss the various aspects of mixing sugar in materials and ask them to tell you what they "know"

about what happens when you add something like sugar or salt to a liquid. Write these on the "Best Thinking So Far" chart for discussion. Many children will say that the sugar or salt disappears. After talking about safety procedures, which would include the rule that they should not taste anything unless an adult says it is okay, ask them if they can think of a way to see if the sugar really disappears into a liquid. Have students put sugar into paper cups filled with warm water at their desks and observe what happens to the sugar. You can say that it is okay to taste the liquid and allow them to take a sip of the solution. They will taste the sugar so that you can ask them if the sugar really disappeared. The same thing can be done by adding a bit of lemon juice to the solution and asking them to taste that to see if the lemon juice disappeared. When they say they can taste both the sugar and the lemon, you can introduce the term *solution* and then continue to refer to solutions as you proceed through the lessons and the inquiries. You might also introduce the term *dissolve* here and tell them that the sugar and lemon dissolved in the water.

Some children may say that the sugar melted in the water. You can tell them that melting happens when you add heat to a solid and that it becomes a liquid. Showing them an ice cube and a sugar cube and allowing the ice cube to melt will be an example of melting. The sugar cube, of course, will not melt at room temperature. Putting the sugar cube into water will be an example of dissolving and the difference can be stressed. You might then make a chart that has two columns, "Melting" and "Dissolving," and ask the children to place events they have experienced in the proper column. Then let them discuss the columns and agree or disagree with the things that have been written. Ask them to use their knowledge of what constitutes melting and dissolving to decide if the event belongs in one column or another. Having them discuss as a class is very important here since entering into a discourse on the topic allows students to listen and respond to one another using evidence from their experiences.

Of course, reproducing the events of the story is in order and if your students do not suggest temperature as a variable, you may have to entice them to try that. Using syrup or making syrup is another way to look at Caroline's problem. Caroline could have won the bet had she made her sugar into syrup and added that to the iced tea. Next time she will know better—after students give her a clue!

You may also find the article by Peggy Ashbrook "Mixing and Making Changes" from *Science and Children* (2006) helpful since it looks at these concepts from an early childhood level.

USING THE STORY WITH GRADES 5–8

After listening to the story, a lively discussion should ensue since the narrative involves a bet and a challenge. The "Our Best Thinking" chart could include suggestions for Caroline to dissolve sugar in the tea.

Experience tells us that most suggestions will involve heat. However, sometimes students suggest grinding the sugar up into smaller particles or using powdered sugar. These students may realize that the smaller the particle, the more surface area is in contact with the water and the less the particles have to be broken down into the molecular level. Some will suggest making syrup out of the sugar and a little water, their experience telling them that liquids dissolve in liquids more easily than solids in liquids. In syrup, the sugar molecules have already been broken down into smaller molecular sizes and will mix with the water more quickly.

There are obviously investigations to be done here. All of the above suggestions can be changed into questions and therefore into investigations. One caution about investigations of this sort is in order: Students need to define what they mean by "dissolve." How will they decide when the substance has dissolved? Two different groups may have different criteria and thereby not reach a consensus. I call this making an *operational definition*. They will also have to develop a *procedural definition*, which is a description of how they will add the solute to the solvent and in what amounts.

Many predictions are to be expected, such as

- The hotter the water, the more sugar can be dissolved.
- The hotter the water, the faster sugar will dissolve.
- Some liquids will dissolve more sugar than others.
- Some liquids will dissolve sugar faster than others.
- Syrup will dissolve more sugar in water than solid sugar.
- Syrup will dissolve sugar faster than solid sugar.
- Powdered sugar will dissolve faster than sugar crystals.
- Powdered sugar will dissolve more than sugar crystals.
-

In the end, your students should be able to keep data tables in their science notebooks and also graph the results. See Michael Klentschy's book *Using Science Notebooks in Elementary Classrooms* (2008). Don't let the title dissuade you from using it, because it is valuable for middle school students as well. The students should find out that temperature does make a difference in dissolving amounts and rates; that there is a saturation point for all solutions; that particle size does affect amount and speed of dissolving; and that the sugar can be reclaimed from the solution by evaporating the water either by heat or by natural evaporation. Another activity that might come up is making rock candy by supersaturating some sugar solution and then hanging a string seeded with some sugar crystals into the solution while it cools. The giant crystals that grow on the string will provide more evidence of the presence of sugar in the solution.

Finally, to see a graphic and very understandable explanation of how things dissolve, visit the Northlands Community College Biology Animations website: *www.northland.cc.mn.us/biology/Biology1111/animations/dissolve.html.*

RELATED NSTA PRESS BOOKS AND JOURNAL ARTICLES

Driver, R., A. Squires, P. Rushworth, and V. Wood-Robinson, 1994. *Making sense of secondary science: Research into children's ideas.* London and New York: Routledge Falmer.

Keeley, P. 2005. *Science curriculum topic study: Bridging the gap between standards and practice.* Thousand Oaks, CA: Corwin Press.

Keeley, P. 2008. *Science formative assessment: 75 practical strategies for linking assessment, instruction, and learning,* Thousand Oaks, CA: Corwin Press.

Keeley, P., F. Eberle, and L. Farrin. 2005. *Uncovering student ideas in science: 25 formative assessment probes* (vol. 1). Arlington, VA: NSTA Press.

Keeley, P., F. Eberle, and J. Tugel. 2007. *Uncovering student ideas in science: 25 more formative assessment probes* (vol. 2). Arlington, VA: NSTA Press.

Keeley, P., F. Eberle, and C. Dorsey. 2008. *Uncovering student ideas in science: Another 25 formative assessment probes* (vol. 3). Arlington, VA: NSTA Press.

Klentschy, M. 2008. *Using science notebooks in elementary classrooms.* Arlington, VA: NSTA Press.

references

American Association for the Advancement of Science (AAAS). 1993. *Benchmarks for science literacy.* New York: Oxford University Press.

Ashbrook, P. 2006. Mixing and making changes. *Science and Children* 43 (5): 28–31.

Keeley, P., F. Eberle, and L. Farrin. 2005. *Uncovering student ideas in science: 25 formative assessment probes* (vol. 1). Arlington, VA: NSTA Press.

Klentschy, M. 2008. *Using science notebooks in elementary classrooms.* Arlington, VA: NSTA Press.

National Research Council (NRC). 1996. *National science education standards.* Washington, DC: National Academy Press.

Northland Community College, Minnesota. Animations website. *www.northland. cc.mn.us/biology/Biology1111/animations/dissolve.html*

CHAPTER 18
COLOR THIEVES

Jenny liked to do projects for school. Well, actually what she really loved to do best was decorate the covers on her projects. She loved colors and shapes and would spend as much time on the covers as she did on the rest of the project.

This particular day, Jenny was just finishing a project for science. Balloons were the topic—helium and hot air balloons and what causes certain balloons to rise up and float away. The hot air balloons were her favor-

ites, maybe because they were so colorfully decorated with big flashy designs and shapes. Jenny lived in a valley where hot air balloons were in the air almost every morning and evening. You could hear the roar of the flame as the pilot made the balloon go higher.

The project was interesting and Jenny worked hard explaining what she had found out about how these big bags of hot air could float for long distances all over the countryside. But now came the fun part, creating the cover sheet and making it colorful and bright.

She had little trouble deciding on what she would draw.

"I'm going to draw a big hot air balloon with a star pattern made up of green, purple and blue diamonds on it," she told her mother. Then she set about drawing her patterns and coloring them in.

"I've got some beautiful see-through colored plastic covers," said Mom. "You can put your project in one of those and make it really special and colorful."

"Cool," said Jenny and continued to color in the reds, yellows, blues, purples, and greens.

It was spectacular! It was a masterpiece! Jenny was proud of both the project and the cover, but mostly the cover.

"Mom!" she shouted, "Where are those colored folders you told me about?"

"I put them on the dining room table," Mom called from the living room. "Take your pick. I think there are five or six different colors."

There were. Green, red, yellow, blue, green-blue, and colorless folders sat on the table. Which one to use? Jenny pondered. How about red? That should make all of the colors look brighter.

She slid the project cover under the red see-through folder and stepped closer to the table lamp for a good view. It looked…

"Gross!!" Jenny shouted. "The cover stole some of my colors!"

Her mother came in to see what the fuss was about.

"What do you mean, 'stole your colors'?"

"Look," said Jenny and held up the folder and project.

"You're right!" exclaimed Mom. "Some of the colors are… BLACK and others are gone! How can that be? I thought the red would make the colors all seem, well, prettier. Let's try the other covers."

And they did and the results were even more surprising. Later, Jenny slid the project into the clear folder and admired her work. But she still wondered how the other folders managed to steal her colors. And she wondered if she could ever make a cover that would not change even if it were put in a colored folder.

NATIONAL SCIENCE TEACHERS ASSOCIATION

PURPOSE

This story poses a challenge to its readers to solve the mystery of light, color, and how we see color. It also asks the question, "What is color?" After investigating the phenomena of color and color filters, students should realize that light is made up of many colors in our visible spectrum and our eyes and brain contribute to the process of "seeing" color.

There is a secondary and broader purpose in this story as well: Help students understand that we see things, no matter how shiny or dull they are, by light that is reflected from the objects to our eyes. This is one of the most difficult concepts to teach. I hope that this story and the activities it generates will lead to at least a more enlightened view of how we see objects.

DON'T BE SURPRISED

Most children and many adults believe that sight originates in the eye. It is said that beauty lies in the eye of the beholder, but that is *not* true of sight. We see objects because light reflects from objects into our eyes. So without light, we simply cannot see. Most of your students will believe that, given enough time for our eyes to adjust, we can see in total darkness. Thus, the idea of seeing because of reflected light is counterintuitive. In the DVD *A Private Universe* produced by the Harvard-Smithsonian Science Media Group, Karen, an eighth grader, sits in a completely dark room for 15 minutes and swears that she sees the yellow side of a block on the table in front of her. When the interviewer showed her that the yellow side of the block had been turned away from her, she still insisted that, given enough time, her eyes would eventually adjust. Despite the complete lack of light, she would be able to see at least the outline of the block.

Students also do not believe that light travels from place to place, but that light merely "fills" a space such as a room. Since light travels so fast, it is not hard to see how this misconception formed. Probes ("Can It Reflect Light?") and ("Apple in the Dark") from *Uncovering Student Ideas in Science*, volume 1 (Keeley, Eberle, and Farrin 2005) are aimed at finding students' ideas about light.

RELATED CONCEPTS

- Energy spectrum
- Color
- Reflection
- Diffraction
- Mixing colors
- Primary colors

CONTENT BACKGROUND

Before we begin, I must emphasize that our perception of color is just that—we *perceive* color by the functions of our eyes, our brains, and what we have been taught. In other words, the light and color we "see" is a combination of learning and the capacity of certain cells in the eye, which in turn communicate with the brain. What we perceive seems "real" to us, but that perception may give us some false beliefs about the true nature of light. It can also explain why we all have resistant misconceptions about color and sight.

The behavior of light is very complex and I cannot possibly explain all of the intricacies in this volume. If you wish a more formal, detailed explanation about the various waves, dig into your copy of *Science Matters* and find the areas that are still confusing to you (Hazen and Trefil 1991). A very fine explanation covering many pages can be found in the teachers' section of the GEMS curriculum booklet *Color Analyzers,* from the vast number of units published by the Lawrence Hall of Science (Erickson and Willard 2005). You can also get a copy of Bill Robertson's *Light: Stop Faking It! Finally Understanding Science So You Can Teach It* (2003). This book has built-in activities that will help you understand light like you never have before. It is not necessary that you have a graduate degree in physics to teach this story but a little background will help you when you run up against your students' data and questions. It will help you ask the right questions or deliver the right comment to send your students back to their investigations with new insight. It will also help you help your students look without telling them what to see.

This story is concerned with the behavior of light and color. Light is the result of an electromagnetic disturbance, and for our purposes we will consider these disturbances as producing electromagnetic *waves.* This is very much like throwing a stone in a pond and seeing the waves or ripples moving outward from the point of contact between the stone and the water. We can compare these with sound waves caused by vibrating objects causing a disturbance in the air molecules around them. The source of the disturbance with light waves is the Sun and the various means for producing light that humans have come up with over the millennia include fire, electricity, battery-powered flashlights, and candles, to name a few.

Light waves have different *wavelengths* (the length between the crests or troughs of two adjacent waves). Humans have named the light of certain visible wavelengths. These are red, orange, yellow, green, blue, indigo, and violet (ROYGBIV). These waves are in our limited range of visible light that we have learned to call *white light.*

ROYGBIV (pronounced like a name: "Roy G. Biv") is an acronym to remember all of the light colors from the Sun in the order that they appear in a rainbow. Rainbows happen when white light is broken down into its component parts by another phenomenon (diffraction). The visible spectrum is a tiny part of the entire family of wavelengths that also includes many to which our senses are not privy. These include infrared, ultraviolet, radio waves, gamma rays, and x-rays.

Because of their various wavelengths and energy levels, visible light waves are prone to being separated by such things as water droplets, prisms, and a device know as a "diffraction grating" (more about this below). This bending and spreading out of the waves is called *diffraction.* In the visible spectrum, each wavelength bends differently so that they all separate into what we know as the color spectrum. In addition, whenever waves enter a medium of a different density, like water or glass, they tend to bend, some more than others and so they spread out. This is another example of diffraction.

Our eyes have millions of color-sensitive cells in the retina called cone cells because of their shape. There are also cells shaped like rods, but these only differentiate between light and dark. Cone cells are activated only in bright light so we do not see color in dim light such as moonlight. It is believed that there are three types of cone cells, each of which has sensitivity to light of different wavelengths. These sensitivities are attuned to the wavelengths we call red, green, and blue. These are often called *primary colors* of light because when they are combined in equal intensities, we perceive white light. Please note that I said we *perceive* white light. This is important because, as I stated before, color is all about perception. Red, green, and blue light can also be combined in various other ways to produce almost any color. Your color TV set or computer monitor does this with the three kinds of pixels to produce the multicolor images you see on the screen.

"Wait!" you say. "Stop the presses! Misprint? Yellow, not green, is a primary color!" True, yellow is a primary color in *pigment* colors but not in *light* colors. We have to concentrate on the *light colors* to understand why things have the colors we see. This is an extremely important distinction. In writing this content section for you, I confess that after all these years, *I* finally understand why we see color! Basically it all boils down to concentrating on these primary *light colors*. Remember this vital point as we move along.

Since visible light is really made up of the various colors of the rainbow (ROYGBIV), when it strikes certain objects the objects absorb some of the colors and reflect others. The colors that are reflected give that object what we perceive as its color. For example, we see green leaves as green because leaves reflect the green wavelengths back to our eyes and then to our brain, which tells us, "Those leaves are green." We know this because way back in our childhood, someone taught us that when we see, that particular color, we call it "green." If, we were German we would call it *grün*, if we were Spanish, we would call it *verde*, and if French, *vert*. In other words, color names are a human construction.

All other colors in the white light are absorbed by the leaf and either warm the leaf up or help in the photosynthesis process. So that is why the leaf appears green—it reflects green light waves. By the same token, green paint is green because it reflects green light waves and absorbs all others. In other words, the colors of an object as you see it is not *in* the object itself but in the colors that the object selectively absorbs and reflects. White is not a color but the combined wavelengths of the visible spectrum. Truly white surfaces should reflect the entire visible light spectrum. This may not be true if you paint your walls with "Cirrus" or "Summer Cloud" or any other fancy name for an off-white color bought at the paint store.

Objects appear to be different colors when seen in different light environments. You may have taken a paint or fabric sample out of the fluorescent lighting in a store to look at it in the sunlight. What you see is the reflection of the elements of the light spectrum produced by the light source you are using. Not all light bulbs produce the same spectrum and this will affect what the objects reflect. The new energy-saving bulbs have a different spectrum than the traditional incandescent bulbs, so their colors appear to be different. Things look different at dusk, as the Sun gets lower in the sky, than they do at noon.

Let's look for a moment at the difference between the mixing of colors as light and the mixing of pigments. Mixing of colors of light is called an *additive* process and mixing of pigments is a *subtractive* process. That needs some explanation!

First, we should look briefly at the additive process. It helps if you imagine everything dark and think that you are now going to add primary light colors to that darkness. Primary colors are defined as colors that when combined produce white *if they are colors of light* and produce black or gray *if they are pigments*. Thus, the three primary additive colors of light are red, green, and blue.

In the additive process, we combine the primary light colors R (red), B (blue), and G (green) to make W (white). Let's create a formula that says R + B + G = W. Now sometimes we mix less than all three. Let's say we mix R (red) + B (blue). That would appear as M (magenta). If we mix B (blue) + G (green), we get C (cyan), akin to aqua; and R (red) + G (green) = Y (yellow). This leads into a whole other set of color bases in paint, but we won't go there right now. Colors of light and pigment are challenging enough!

I held up a yellow notebook and suddenly realized it appeared yellow because it absorbed the B (blue) and reflected the R (red) and G (green) to my eyes. As we can see from above, R + G = Y! So the color of an object is not *in* the object but in the primary light color(s) it absorbs and the remainder of the spectrum it reflects to our eyes! In a formula it would look like:

$$(R + B + G) - B = R + G$$

$$R + G = Y$$

Written out, it would be: Red plus blue plus green (which combine to make white) minus blue equals red plus green, which equals yellow. So, that's why I saw yellow! I saw the combination of the reflected red and green that in light make yellow. Or it may be even a little more complicated than this, because it is possible that yellow light was reflected from the spectrum as well. My eye however, was stimulated to see yellow either way. I could not distinguish the difference. It was all yellow to me. Thus perception of color is once again related to how our brains translate what we see.

I must be clear that the whole theory of color perception is still in the process of study. What I have tried to explain to you is close to the accepted theory at this point. This may be all too complicated for elementary school kids to understand but I hope it makes things clearer to you and it might well be appropriate for middle school.

Next, consider mixing pigments, the subtractive process. If you have a blue pigment, you see it as blue because only blue is reflected from its surface. If you mix this blue pigment with yellow, the mixture will seem green but *not* yellow or blue. Thus you have *subtracted* from your vision two colors that you could have seen separately, but cannot see combined. Less light is reflected because you have

NATIONAL SCIENCE TEACHERS ASSOCIATION

put together two colors that are going to absorb more light, thus *subtracting* the number of different light waves reflected to your eyes. The more pigments you add, the more light will be absorbed and the less reflected until so little light is reflected that you perceive it as gray or black. I'll bet you all have had some experience with primary colors in pigments. A favorite activity is to take white frosting from the can and give kids food coloring to make different color frostings to go on graham crackers. Not the healthiest activity, but once in a while it shouldn't hurt (except for diabetic or sugar-intolerant children).

This helps us understand a little more so we can talk about filters. What do colored filters do? One thing they do *not* do is change the color of light. Colored filters let certain colors through and do not let other colors through. For example, a red filter allows red, yellow, orange, and pink through to varying extents (which blend in with a white background). On the other hand, brown, green, blue, and purple appear black because the light that they reflect cannot penetrate the red filter. Their wavelengths are completely absorbed, thus appearing black.

When you look at objects through a red filter, that filter only lets red or near-red colors get to your eyes. It absorbs green and blues. Why do I say *near*-red colors? Yellow pigments from crayons, markers, and other paints often reflect some red and green light as well as yellow. It is no wonder yellow objects seem so bright since they are reflecting more different colors. Can you guess why many emergency vehicles are painted yellow?

Orange objects also reflect some yellow and red as well as orange. Orange appears bright through a red filter. Do you think this is true if seen through a green filter? Give yourself points if you said no since the green filter would filter out yellow and red as well as the orange, since first, it absorbs red, and then you need red and green to make yellow. Red objects seen through a red filter will seem to disappear into the white background because the filter only allows red light from the background. So, the red object seems to fade into the background, sort of like camouflage.

Your students will be finding these things out as they make drawings and notes using different crayons and markers, then viewing them through red and green filters. There will be more details in the "Using the Story" sections. I suggest that you make yourself a pair of red filter glasses, get some colored markers, and try these things as you read this material.

I see this as an exciting chance to show that we see only that from which light is reflected. If light were not reflected from the things we see through the filter, the filter would not alter them! Perhaps this will convince some and perhaps not. But it certainly will not hinder their growth toward understanding light, color, and electromagnetic energy, if not now, maybe at another time in the future.

related Ideas From National Science education Standards (NrC 1996)

K–4: *Light, Heat, Electricity, and Magnetism*
- Light travels in a straight line until it strikes an object. Light can be reflected by a mirror, refracted by a lens, or absorbed by the object.

5–8: *Transfer of Energy*
- Light interacts with matter by transmission (including refraction), absorption, or scattering (including reflection). To see an object, light from that object—emitted by or scattered from it—must enter the eye.
- The Sun's energy arrives as light with a range of wavelengths, consisting of visible light, infrared and ultraviolet radiation.

related Ideas In BenCHMarkS For SCIenCe LITeraCY (aaaS 1993)

6–8: *Motion*
- Light from the Sun is made up of a mixture of many different colors of light, even though to the eye the light looks almost white. Other things that give off or reflect light have a different mix of colors.
- Something can be "seen" when light waves emitted or reflected by it enter the eye—just as something can be "heard" when sound waves from it enter the ear.
- Human eyes respond to only a narrow range of wavelengths of electromagnetic radiation—visible light. Differences of wavelength within that range are perceived as differences in color.

USING THe STOrY WITH Grades K–4

Although this is one of my very favorite stories, I admit that I was a bit hesitant to recommend it below grade 6. But then I read "Secret Message Science Goggles" by Christina DeVita and Sarah Ruppert in *Science and Children* (2007) and was convinced that it could be used at least as low as second grade and perhaps in earlier grades as well. Devita and Ruppert were not focused on the physics of the topic but tried to provide the students with a tool with which they could investigate the

topic of color. I would suggest that you give your class the probe(s) "Can It Reflect Light" and "Apple in the Dark" from *Uncovering Student Ideas in Science*, volume 1 (Keeley, Eberle, and Farrin 2005). These probes will give you a head's up on what your students believe.

You might want to create two "Our Best Thinking" charts on "What We Know About Seeing" and "What We Know About Color." After the discussion, you can have the children make goggles with red filters out of cardboard (glue or tape transparent red plastic film over each eye hole). Have them describe and record their findings as they look around the room. They will be seeing the world through a filter that allows only red light to reach their eyes. Let them explore the world this way for a while and raise questions about what they observe on objects around the room, patterns on clothing, and pictures on the walls. You can write some secret messages, or put a red bird in a green cage, which will disappear or at least look ghostly when viewed through the red goggles.

Return to the charts and begin to develop investigations about color and seeing by using the goggles or perhaps making others with blue or yellow filters. Students usually want to re-create the situation in the story and make a picture with many colors to be viewed through filters. This is a good idea since it puts the students in Jenny's position. They feel like they are "in" the story.

Your students will notice that certain colors will look black as seen through a red filter. Suppose they look at a green book cover or marker swatch. The red filter blocks out blue and yellow so that they do not penetrate the filter. This means that since there is no red in the green cover, nothing is transmitted through the filter and the cover looks black. Would the same be true if the cover were blue? Of course, since no color would penetrate the filter, the image of the cover would still appear black. Students will figure out what colors appear black and which they will be able to see and that if Jenny had avoided these colors she could have avoided the blackouts in her cover. They will also be able to see that by using different filters, they can hide words and other messages in designs and send secret codes by use of various filters.

Finally, the students are left with a challenge from the last line of the story. Is it possible to make a cover with colors that will not change if they are put in a colored folder? This can provide not only a great discussion but give the students an opportunity to try their ideas and get instant feedback. It will also give them a chance to finish the story.

USING THe STORY WITH GraDES 5–8

Students at this age also like to "live the story" with Jenny just to get the feel of the situation and see if it really is true. This means that you need to have a few see-through folders of different colors around for the students to try. Just like the younger children in the prior section, they can be encouraged to write secret messages that can only be seen through a filter.

Students also respond very well to the probes mentioned in the prior section and get into a lively discourse about whether things reflect light and whether or not you can see in total darkness. I might add that most children have never experienced total darkness unless they have been in a cave or someplace similar, so their experience is usually *semi*-darkness. They will be adamant that in time their eyes will adjust. I have yet to visit a school where achieving total darkness is possible. If your school is lucky enough to have a photography darkroom, maybe that would work.

Your students will likely insist that shiny things like mirrors or windows will reflect light but not rocks or other dull objects. This is why "Color Thieves" is a story that might just provide them with enough disequilibrium to question their prior conceptions. First it will be necessary for you to show them that white light is made up of many colors. This means that you will have to have a sheet of diffraction grating so that students can view light sources and see the various spectra that make up the light. Diffraction gratings can be purchased from online distributors such as Edmund Scientifics or Delta Educational. They are sheets of plastic with thousands of lines scribed into them so that light will be scattered and produce a spectrum. If the children bring old CDs from home they can tilt them and see a spectrum on them as well due to the diffraction of light on the surface of the CD. If you follow the ideas suggested in the GEMS unit *Color Analyzers* you can make up what they call color analyzers: index cards with three holes. One hole has a red filter, another has a green filter, and the third hole has the diffraction grating cut from a larger sheet. This saves money since the amount of filters and gratings used are small.

Using several types of lamps is helpful if the students are looking for differences in kinds of spectrums through the diffraction grating. Different hues may be missing or of varying brightness. This is another form of evidence that the colors we see are perceptions. Fluorescent lamps used to be rather narrow but now are manufactured to have a broader spectrum. If you have some older fluorescent lamps to compare to modern ones, this is an interesting activity.

Filters can be purchased at most art stores for very little money. The quality varies a great deal, which sometimes affects the results, but most of the time the main colors come through as expected. The folks at Lawrence Hall, the GEMS people, recommend the use of Crayola Classic brand markers for the truest colors. Take along a paper with color marks on it to help you test the filters you want to purchase.

You may want to return to the story and see that Jenny drew her cover using colors such as reds, greens, purples, blues, and yellows. Jenny's reds may have "disappeared" into the background of the white paper; certainly the greens and purples and blues would have seemed black and the yellow might also have blended into the background. If this is the case, then students can write or draw on paper with different colors and then look at their papers through filters and find completely different messages or patterns. Certain color letters mixed in with other color letters might seem to say one thing without the filter and another through the filter. They will love to make secret messages.

Finally, the last sentence in the story does offer a challenge. I recommend a good discussion before letting the children try to create such a paper. Remember, differences of opinion are the key to good discourse and also let you in on imbedded assessment as you listen to the discourse. Here you might interject a few questions that will lead the children toward wondering if there isn't evidence here that shows that light has to be reflected from an object in order for them to see that object.

related NSTa Press Books and Journal articles

DeVita, C., and S. Ruppert. 2007. Secret message science goggles. *Science and Children* 44 (7): 30–35.

Driver, R., A. Squires, P. Rushworth, and V. Wood-Robinson. 1994. *Making sense of secondary science: Research into children's ideas.* London and New York: Routledge Falmer.

Keeley, P. 2005. *Science curriculum topic study: Bridging the gap between standards and practice.* Thousand Oaks, CA: Corwin Press.

Keeley, P., F. Eberle, and L. Farrin. 2005. *Uncovering student ideas in science: 25 formative assessment probes* (vol. 1). Arlington, VA: NSTA Press.

Keeley, P., F. Eberle, and J. Tugel. 2007. *Uncovering student ideas in science: 25 more formative assessment probes* (vol. 2). Arlington, VA: NSTA Press.

Keeley, P., F. Eberle, and C. Dorsey. 2008. *Uncovering student ideas in science: Another 25 formative assessment probes* (vol. 3). Arlington, VA: NSTA Press.

Klentschy, M. 2008. *Using science notebooks in elementary classrooms.* Arlington, VA: NSTA Press.

references

American Association for the Advancement of Science (AAAS). 1993. *Benchmarks for science literacy.* New York: Oxford University Press.

Astrophysics Science Media Group. 1995. *A Private Universe,* DVD.

DeVita, C., and S. Ruppert. 2007. Secret message science goggles. *Science and Children* 44 (7): 30–35.

Erickson, J., and C. Willard. 2005. *Color analyzers: Investigating Light and color.* Berkeley, CA: GEMS, Lawrence Hall of Science, University of California.

Hazen, R. and J. Trefil, 1991. *Science matters: Achieving scientific literacy.* New York: Anchor Books.

Keeley, P., F. Eberle, and L. Farrin. 2005. *Uncovering student ideas in science: 25 formative assessment probes* (vol. 1). Arlington, VA: NSTA Press.

National Research Council (NRC). 1996. *National science education standards.* Washington, DC: National Academy Press.

Robertson, B. 2003. Light: *Stop faking it! finally understanding science so you can teach it.* Arlington, VA: NSTA Press.

A MIRROR BIG ENOUGH

Tenika needed a mirror for her room. She had a few problems to solve before she bought one, though. She had a very small room and finding a place for a full-length mirror was not easy. Naturally she wanted to see her image from head to toe, and she was tall for her age. There was only one wall that would hold a full-length mirror unless she wanted to rearrange all of her bedroom furniture, which she didn't. She had to learn a lot more about mirrors before she made her purchase. She also had a limited amount of money, so that created another problem.

At the furniture store Tenika looked all around and fell in love with a freestanding walnut mirror that would be just right. If she bought that one, she could put it anywhere in the room and the problem was solved. That was, until she looked at the price tag. Three hundred dollars! Well, that was that.

All of the other full-length mirrors that she liked were expensive too. There were a lot of cheaper and smaller mirrors for sale but she still had her heart set on seeing all of herself, not just part of her. All of the small mirrors were mounted on the wall just at her eye level so she could see her face from hair to chin clearly. But there was the problem of the size of her room. Could she get a smaller mirror and stand far enough away to see all of her? How far back did she have to move to see all of her in a small mirror? As she tried moving back from each mirror, she found that what she saw was not what she expected.

"What is going on here?" she wondered. "Are these mirrors weird, or what?"

She wandered around and tried looking in all of the mirrors. She found out a few things about how mirrors work that surprised and amazed her. Also, some of the mirrors were mounted too high and some were too low and that seemed to make a difference in what she could see.

Since the smallest mirrors were the cheapest, Tanika kept asking herself, What is the smallest mirror I can buy so that I can still see myself, head to toe? In the end, she was able to buy a mirror she could afford that worked for her.

PURPOSE

Mirrors are part of our everyday lives. How many of us start our day in front of the bathroom mirror brushing our teeth or our hair? Our early ancestors probably did much of the same by looking into pools of water and seeing their reflections. And it does not matter how supple you think you are and how much experience you have had with yoga; it is still impossible to see the back of your head without the use of two mirrors. This story is designed to motivate students to explore how mirrors work and how mirrors reflect the light (first reflected from objects to the mirror and then to our eyes as images), and to discover in this particular case the famed rule that "the angle of incidence equals the angle of reflection." Think of the many times a mirror is used in your daily life and think also how much *you* understand how it works like it does. Mirrors are indeed the source of some of the most illusive everyday science mysteries.

DON'T BE SURPRISED

Children and adults alike tend to believe that mirrors and shiny things are the only objects that reflect light. Actually it is only because everything reflects light that we are able to see anything. If you give the probe "Can It Reflect Light?" from volume 1 of *Uncovering Student Ideas* (Keeley, Eberle, and Farrin 2005), you will find most of your students will probably not realize that all objects reflect light in different ways so that we *see* them as different objects. Although it is not absolutely necessary for students to believe that everything reflects light to their eyes in order to understand how mirrors work, it gives you as a teacher insight into the ways your students perceive the concepts of light and sight. In fact, it is one of the most difficult of preconceptions to change. I believe that the topic explored in the previous chapter "Color Thieves" stands the best chance of modifying this preconception, particularly for older students. But that is a different story altogether (literally!).

Adults also believe that they do not have to stand directly in front of a mirror in order to see their own image. Most believe that they can stand at the edge of a mirror and see their likeness. It is really strange that something we use everyday is still an enigma and the source of so many misconceptions.

RELATED CONCEPTS

- Light energy
- Reflection
- Vision
- Mathematics (angles)

CONTENT BACKGROUND

The image we see in a mirror is not real. It is a *virtual image,* or what we have learned to call a *mirror image,* inside the mirror or behind it. A mirror is a piece of glass that has aluminum or silver backing bonded to it. It is important for the metal to be present because when light energy reflected from an object in front of the mirror hits the metal backing, it excites the electrons in the metal. They respond by giving off light energy directly back in the same direction from which the light originated. Mirrors are very smooth and polished so that the light that is reflected back is barely distorted and gives a fairly accurate image of the object placed in front of it. Historically, mirrors were probably polished metals. Mirrors as we know them today were probably not available until the 18th or 19th century. As early as the 12th century, metal-backed mirrors were used in Europe; before that time, silver or bronze polished to high brilliance was used. These mirrors were still not as good as the mirrors of today.

I have mentioned that the image you see is not real. It cannot be touched and it has no mass, nor does it take up space. In essence, it is another dimension. It is really behind the surface of the mirror, as you perceive it. It is also twice the distance behind the mirror as you are in front of the mirror. Yet your image in the mirror is the same size. Think about that for a second.

Light is reflected from mirrors directly back toward the direction from which it came, if you are standing directly in front of the mirror. That is, if light strikes a mirror head on, the light is reflected directly back. If the light from the object strikes the mirror at an angle, the image is reflected back at that same angle but in the opposite direction. This is very much like a ball bouncing off a wall. If you throw the ball directly at the wall, it comes right back to you. But, if you throw it at an angle, it will bounce off the wall and go off at exactly the same angle in the opposite direction. The rule is that the angle of incidence (the angle that the ball hits the wall) will equal the angle of reflection (the angle the ball will move when it leaves the wall).

This is why you can see images of objects in a mirror even though you are viewing from a point to one side or the other. If you can see a person in a mirror, they can see you as well. You may have seen trucks or buses with a safety sign on the back that says, "If you can't see my mirrors, I can't see you." This means that if you can see the truck's mirrors, the reflection from you is visible in the truck's mirror as well.

You may have noticed that when you look at your side mirrors while driving or while sitting in the passenger seat, you do not see yourself but only the road behind you. You are at an angle outside the reflection angle of the mirror because it is set to see what is behind the car. The same is true of the inside rearview mirror, which is set to see out of the rear window. If you were to look at the rearview mirror through the rear window of the car you would be able to see person in the driver's seat. Try it.

Another conception common among children and adults is that if you move back from a mirror, you will see more of yourself. This preconception probably has its origins in the bathroom where we do most of our mirror gazing. Because most bathroom mirrors are mounted on the wall in front of a counter and cabinet, we are deprived of any reflection of objects below the level of the counter. As we move back from the counter, more of us can be reflected into the mirror because the counter no longer prevents it and we see more of ourselves, particularly that part below the waist. So, as long as your eyes are in the same horizontal plane as the mirror on the wall and the mirror

is flat against the wall, you will see the same image whether you are close to or further away from that mirror.

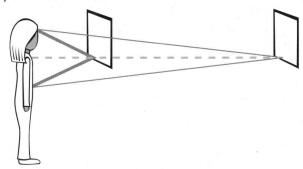

Now, let us imagine a mirror that is half the height of the girl in the picture. As she looks at the mirror, the angle of the reflected light from her feet will equal the angle that meets her eyes. Therefore, if the mirror is half the height of the person looking into it and the mirror is in the same horizontal plane as that person's eyes, she can see her whole body.

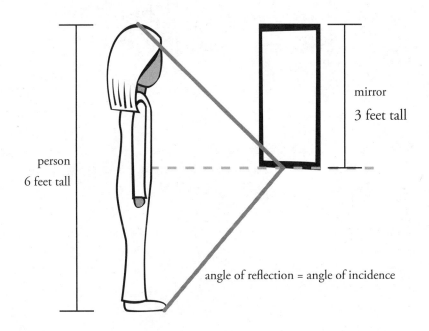

person
6 feet tall

mirror
3 feet tall

angle of reflection = angle of incidence

This will explain how Tenika solved her problem. She had to get a mirror at least one half her height, but considering that she would be growing, she had to estimate how tall she would be and buy a mirror half that size. She also realized that it made no difference how far she was from the mirror; she would see an image of herself that showed the same amount of her body.

To obtain some information on optical properties of light I refer you to *Science Matters* by Hazen and Trefil (1991) pages 106–108.

related ideas from National Science education standards (NrC 1996)

K–4: Light, Heat, Electricity and Magnetism
- Light travels in a straight line until it strikes an object. Light can be reflected by a mirror, refracted by a lens, or absorbed by the object.

5–8: Transfer of Energy
- Light interacts with matter by transmission (including refraction). To see an object, light from that object—either emitted by or scattered from it—must enter the eye.

related ideas in Benchmarks for Science Literacy (aaas 1993)

3–5: Motion
- Light travels and tends to maintain its direction of motion until it interacts with an object or a material.
- Light can be absorbed, redirected, bounced back, or allowed to pass through. (Note: This is a new benchmark. It can be found in AAAS 2001, p.63.)

6–8: Motion
- Something can be "seen" when light waves emitted or reflected by it enter the eye.
- Light acts like a wave in many ways. Waves can explain how light behaves. (Note: This is a new benchmark. It can be found in AAAS 2001, p.65)

USING THE STORY WITH GRADES K–4

Although the Standards and particularly the Benchmarks do not have much to say about this concept in the early years, the story can at least be a grabber to get the kids interested in trying to find out things about mirrors. Young children are marvels at trying new things with objects. With a little encouragement, you should have them telling you all sorts of things they "know" about mirrors. These

NATIONAL SCIENCE TEACHERS ASSOCIATION

statements, of course, can be put on the "Our Best Thinking" chart, changed to questions, and then tested. There is a great deal to be done with younger students and mirrors. To be sure, the young child may not have the vocabulary or the developmental acuity to understand all of the ideas behind the concept of reflection but their experiences with these objects are nonetheless valuable.

Kindergarteners and first and second graders love to play with mirrors and with guidance can learn a lot about the direction of the beams of light through such games as "mirrors and flashlights," where a child is asked to shine her light on her mirror and make the light shine on a specific object in the darkened room. For example, "Sally, make your mirror make the light shine on the clock." The child has to hold the light and the mirror in such a way that the light beam is reflected to that specific object. Then the child explains how she did it. In essence this exploration of mirrors and light sources is a form of inquiry especially if they are asked to predict why they are positioning their objects to get a desired result. They can draw these trials in their science notebooks and explain what happened and possibly try to explain why they think it did. These notes can be of great assistance to you in assessing where to go next.

They can play a mirror-acting game where two children face each other, and they pretend there is a mirror between them. One child acts as the leader and makes an action and the other has to respond or mimic the actions of the first child as though she was the mirror image. Then they switch roles. They can talk about what adaptations they had to make to move just like the leader. For example, "When you raised your right arm, I had to raise my left arm 'cause I'm in the mirror."

USING THE STORY WITH GRADES 5–8

With students from grades 3 or 4 through 8, I suggest starting the lesson by giving the probe "Mirror on the Wall" from *Understanding Student Ideas in Science*, volume 3 (Keeley, Eberle, and Dorsey 2008). This probe asks the students to choose from three options about a girl looking in a mirror and stepping back. It asks whether she will see the same amount of her image, less, or more. It addresses one the main preconceptions about seeing oneself in a mirror. It will provide you with a class profile of how students use ideas about light to explain how we see images in mirrors.

As usual, I would begin with asking the students what they "know" about how we see images in mirrors and writing those down on the "Our Best Thinking" chart as described above in the K–4 section. If you do not have access to the above probe, I suggest taking an idea from Eleanor Duckworth of Harvard Graduate School of Education, who begins a class on science education using mirrors. Place a mirror on the wall and ask two students to show where they would stand so that they can see each other's reflections at the same time. Let the class discuss their predictions with reasons and then let the class experiment with their question on small mirrors placed around the room on the walls. With patience and some guidance from questions by you, it is possible that your students will realize that they have to stand so that the angles respective to the mirror are equal when they are both in view. In other words, they will realize that the angle of incidence will be equal to the angle of reflection (although probably not in that language).

It's helpful to put these findings into drawings that will be easier to explain. Be sure to use the science notebooks as you work your way through this story on mirrors. Make sure students keep all of their drawings and comments and especially tell them to note what *doesn't* work. With your help they can see how when their positions change, the angles change as well.

For other ideas for using mirrors, you can go to the NSTA archives and find the article "Circus of Light" in the February 2004 edition of *Science and Children*. Pay particular attention to the bouncing light section. I prefer when working on this concept reinforcement to have children sit on a floor about six feet from a wall so that they are parallel to the wall in a line. Mark a vertical line on the wall at about the midpoint of the line of children. They can roll the ball so that it hits the line and depending upon the angle at which the ball is rolled, it will roll to a child in the same position as the roller but on the opposite end of the line. Of course it depends upon how accurate the roller is in hitting the mark, but practice makes things go better and children love it. You can then substitute a mirror for the mark on the wall and in a darkened room have each child shine a flashlight on the mirror and the light will end up just as the ball did. This reinforces the idea of angle of reflection and incidence.

Now let's look at the problem inspired by the story. Students are going to try to find out how large a mirror is needed in order to see the whole body. By now they have realized that stepping back is no solution to the problem. The answer must reside in the height of the mirror. In my experience, students suggest using a full-length mirror, mounting it on the wall of the classroom and using paper to cover the bottom part of the mirror. They must predict how much of the mirror would have to be showing so that they could see their whole body. They soon realize that the top of the mirror has to be at eye level to get the best results. Therefore, mounting the mirror does not work since students vary in height enough that the mirror has to be moved to match the eye level of each student.

So, the mirror can be held by several students and moved up or down to find the right spot. Teams of students can work together to take care of all of the tasks involved. Ask them to use their newly found knowledge of reflection to draw diagrams of their vision with arrows and then to predict where the angle that showed their feet would be on the mirror. The angle from their feet to their mirror is the key and that angle has to be equal to the angle that met the eyes. By moving the paper up and down the mirror, they finally can find the spot that just allows them to see their whole body. Again, this varies with students of differing heights but they finally can obtain enough data to realize that the mirror has to be one half of any given student body length to satisfy the need. (See diagram in the background content section of this chapter).

At this point you could distribute the original probe "Mirror on the Wall" that you used as a preinstruction probe. Although all students will probably know that the mirror has to be half the length of a body height, their written explanation will be of more value here in assessing how much they understand what they have discovered. A class discussion of the topic might help clarify the concept for those who are still having trouble understanding the mechanisms of reflection. They are now ready to finish the story with an explanation as well as a solution.

NATIONAL SCIENCE TEACHERS ASSOCIATION

related NSTA Press Books and Journal Articles

Driver, R., A. Squires, P. Rushworth, and V. Wood-Robinson, 1994. *Making sense of secondary science: Research into children's ideas.* London and New York: Routledge Falmer.

Hazen, R., and J. Trefil, 1991. *Science matters: Achieving scientific literacy.* New York: Anchor Books.

Keeley, P. 2005. *Science curriculum topic study: Bridging the gap between standards and practice.* Thousand Oaks, CA: Corwin Press.

Keeley, P., F. Eberle, and J. Tugel. 2007. *Uncovering student ideas in science: 25 more formative assessment probes* (vol. 2). Arlington, VA: NSTA Press.

Keeley, P., F. Eberle, and C. Dorsey. 2008. *Uncovering student ideas in science: Another 25 formative assessment probes* (vol. 3). Arlington, VA: NSTA Press.

Klentschy, M. 2008. *Using science notebooks in elementary classrooms.* Arlington, VA: NSTA Press.

references

American Association for the Advancement of Science (AAAS). 1993. *Benchmarks for science literacy.* New York: Oxford University Press.

American Association for the Advancement of Science (AAAS). 2001. Atlas of Science Literance. Washington, DC: AAAS.

Hazen, R., and J. Trefil, 1991. *Science matters: Achieving scientific literacy.* New York: Anchor Books.

Keeley, P., F. Eberle, and L. Farrin. 2005. *Uncovering student ideas in science: 25 formative assessment probes* (vol. 1). Arlington, VA: NSTA Press.

Keeley, P., F. Eberle, and C. Dorsey. 2008. *Uncovering student ideas in science: Another 25 formative assessment probes* (vol. 3). Arlington, VA: NSTA Press.

Matkins, J. J., and J. McDonnough. Circus of light. *Science and Children* 41(5): 50–54.

National Research Council (NRC). 1996. *National science education standards.* Washington, DC: National Academy Press.

APPENDIX

Some teachers who want to teach using inquiry techniques, and/or teach for conceptual change, like to have a few resources at their disposal. If there is a professional library in your school, the following books would make a fine addition. If not they can be added slowly to your personal library and will soon become dog-eared with use. All these books are available from NSTA (*www.nsta.org/store*).

American Association for the Advancement of Science (AAAS). 1993. *Benchmarks for science literacy.* New York: Oxford University Press.

Driver, R., A. Squires, P. Rushworth, and V. Wood-Robinson. 1994. *Making sense of secondary science: Research into children's ideas.* London and New York: Routledge Falmer.

Hazen, R., and J. Trefil, 1991. *Science matters: Achieving scientific literacy.* New York: Anchor Books.

Keeley, P. 2005. *Science curriculum topic study: Bridging the gap between standards and practice.* Thousand Oaks, CA: Corwin Press.

Keeley, P., F. Eberle, and L. Farrin. 2005. *Uncovering student ideas in science: 25 formative assessment probes* (vol. 1). Arlington, VA: NSTA Press.

Keeley, P., F. Eberle, and J. Tugel. 2007. *Uncovering student ideas in science: 25 more formative assessment probes* (vol. 2). Arlington, VA: NSTA Press.

Keeley, P., F. Eberle, and C. Dorsey. 2008. *Uncovering student ideas in science: Another 25 formative assessment probes* (vol. 3). Arlington, VA: NSTA Press.

Klentschy, M. 2008. *Using science notebooks in elementary classrooms.* Arlington, VA: NSTA Press.

National Research Council (NRC). 1996. *National science education standards.* Washington, DC: National Academy Press.

Each has a different role to play in planning for inquiry teaching. *Making Sense of Secondary Science* is a compendium of research on children's thinking about many science concepts. In this book you will find the kinds of student preconceptions you can expect to be prevalent in your students' minds. *National Science Education Standards* is considered the base upon which all state standard documents are written. This is also true about the *Benchmarks for Science Literacy.* You will also want a copy of your own state's standards. *Science Matters* contains an overview of a broad range of science topics written for popular consumption, clearly stated, and easily understood by the general population. *Science Curriculum Topic Study: Bridging the Gap Between Standards and Practice* does just what the title implies, and *Using Science Notebooks in Elementary Classrooms* will help teachers use this valuable resource for formative and summative assessment.

I wish you all good luck in using the stories to send your students into the realm of inquiry learning. If you have not tried this method before, I sincerely believe that both you and they will notice the difference, and you will all learn science in a very different way. Bon Voyage!

INDEX

Note: Page numbers in **bold** refer to charts or illustrations.